Lectures in Mathematics
ETH Zürich
Department of Mathematics
Research Institute of Mathematics

Alain Valette

Introduction to the Baum-Connes Conjecture

Springer Basel AG

Author's address:

Institut de Mathématiques
Université de Neuchâtel
Rue Emile Argand 11 – CP 2
2007 Neuchâtel
Switzerland

2000 Mathematical Subject Classification 19K35, 20F67, 22D25, 46L80

ISBN 978-3-7643-6706-0 ISBN 978-3-0348-8187-6 (eBook)
DOI 10.1007/978-3-0348-8187-6

A CIP catalogue record for this book is available from the
Library of Congress, Washington D.C., USA

Deutsche Bibliothek Cataloging-in-Publication Data
Valette, Alain:
Introduction to the Baum-Connes conjecture / Alain Valette. - Basel ;
Boston ; Berlin : Birkhäuser, 2002
(Lectures in mathematics : ETH Zürich)
ISBN 3-7643-6706-7

Originally published by Birkhäuser Verlag AG 2002

Member of the BertelsmannSpringer Publishing Group
Printed on acid-free paper produced from chlorine-free pulp. TCF ∞

9 8 7 6 5 4 3 2 1 www.birkhäuser-science.com

Contents

Introduction

A quick description of the conjecture

The *Baum-Connes conjecture* is part of Alain Connes'tantalizing "noncommutative geometry" programme [18]. It is in some sense the most "commutative" part of this programme, since it bridges with classical geometry and topology.

Let Γ be a countable group. The Baum-Connes conjecture identifies two objects associated with Γ, one analytical and one geometrical/topological.

The right-hand side of the conjecture, or analytical side, involves the K-theory of the *reduced C^*-algebra* $C_r^*\Gamma$, which is the C*-algebra generated by Γ in its left regular representation on the Hilbert space $\ell^2(\Gamma)$. The K-theory used here, $K_i(C_r^*\Gamma)$ for $i = 0, 1$, is the usual topological K-theory for Banach algebras, as described e.g. in [85].

The left-hand side of the conjecture, or geometrical/topological side

$$RK_i^\Gamma(\underline{E}\Gamma) \qquad (i = 0, 1),$$

is the Γ-equivariant K-homology with Γ-compact supports of the classifying space $\underline{E}\Gamma$ for proper actions of Γ. If Γ is torsion-free, this is the same as the K-homology (with compact supports) of the classifying space $B\Gamma$ (or $K(\Gamma, 1)$ Eilenberg-Mac Lane space). This can be defined purely homotopically.

The link between both sides of the conjecture is provided by the *analytic assembly map*, or *index map*

$$\mu_i^\Gamma : RK_i^\Gamma(\underline{E}\Gamma) \to K_i(C_r^*\Gamma)$$

($i = 0, 1$). The definition of the assembly map can be traced back to a result of Kasparov [48]: suppose that Z is a proper Γ-compact Γ-manifold endowed with a Γ-invariant elliptic (pseudo-) differential operator D acting on sections of some Γ-vector bundle over Z. Then, in spite of the non-compactness of the manifold Z, the *index* of D has a well-defined meaning as an element of the K-theory $K_i(C_r^*\Gamma)$. On the other hand, using the universal property of $\underline{E}\Gamma$, the manifold Z maps continuously Γ-equivariantly to $\underline{E}\Gamma$, and the pair (Z, D) defines an element of the equivariant K-homology with compact supports $RK_i^\Gamma(\underline{E}\Gamma)$. Then, one sets

$$\mu_i^\Gamma(Z, D) = Index(D).$$

Elaborating on this, and using the concept of abstract elliptic operator (or Kasparov triple), one constructs the assembly map μ_i^Γ, which is a well-defined group homomorphism.

Conjecture 1 (the Baum-Connes conjecture). *For $i = 0, 1$, the assembly map*

$$\mu_i^\Gamma : RK_i^\Gamma(\underline{E}\Gamma) \to K_i(C_r^*\Gamma)$$

is an isomorphism.

This conjecture is part of a more general conjecture (discussed in [9]) where discrete groups are replaced by arbitrary locally compact groups, or even locally compact groupoids: this allows to treat, in a common framework, groups, group actions on locally compact spaces, and foliated manifolds. If one wants to appeal to the powerful techniques of Kasparov's bivariant theory, it is even useful to allow coefficients in an arbitrary auxiliary C*-algebra on which the group(oid) acts; this leads to the Baum-Connes conjecture with coefficients, which computes the K-theory of reduced crossed product C*-algebras.

The reason for restricting to discrete groups is that, in a sense, this case is both interesting and difficult. The difficulty lies on the analytical side: there is no general structure result for the reduced C*-algebra of a discrete group, so that its K-theory is usually quite hard to compute [1]. The interest of Conjecture 1 is that it *implies* several other famous conjectures in topology, geometry and functional analysis.

The Novikov Conjecture. *For closed oriented manifolds with fundamental group Γ, the higher signatures coming from $H^*(\Gamma, \mathbf{Q})$ are oriented homotopy invariants.*

The Novikov conjecture follows from the rational injectivity of μ_i^Γ (see [9], Theorem 7.11; [27], §6).

The Gromov-Lawson-Rosenberg Conjecture (one direction). *If M is a closed spin manifold with fundamental group Γ, and if M is endowed with a metric of positive scalar curvature, then all higher $\hat{A}$-genera (coming from $H^*(\Gamma, \mathbf{Q})$) do vanish.*

This conjecture is also a consequence of the rational injectivity of μ_i^Γ (see [77]).

Let us also mention the conjecture of idempotents for $C_r^*\Gamma$; since $C_r^*\Gamma$ is a completion of the complex group algebra $\mathbf{C}\Gamma$, this conjecture is stronger than the classical conjecture of idempotents, discussed e.g. in [72].

The Conjecture of Idempotents (or the Kaplansky-Kadison conjecture). *Let Γ be a torsion-free group. Then $C_r^*\Gamma$ has no idempotent other than 0 or 1.*

The conjecture of idempotents would follow from the surjectivity of μ_0^Γ (see Proposition 7.16 in [9]; Proposition 3 in [87]).

[1] In many important cases, e.g. lattices in semi-simple Lie groups, the reduced C*-algebra is actually *simple*, see [12].

It has to be emphasized that Conjecture 1 makes $K_i(C_r^*\Gamma)$ computable, at least up to torsion. The reason is that $RK_i^\Gamma(\underline{E}\Gamma)$ is computable up to torsion. Indeed, let $F\Gamma$ be the space of finitely supported complex-valued functions on Γ, with support contained in the set of torsion elements of Γ. Letting Γ act by conjugation on torsion elements, $F\Gamma$ becomes a Γ-module; denote by $H_j(\Gamma, F\Gamma)$ the j-th homology space of Γ with coefficients in $F\Gamma$. In [8], Baum and Connes define a *Chern character*

$$ch_\Gamma : RK_i^\Gamma(\underline{E}\Gamma) \to \bigoplus_{n=0}^{\infty} H_{i+2n}(\Gamma, F\Gamma),$$

and state in Proposition 15.2 of [8] that the Chern character is an isomorphism up to torsion, i.e.

$$ch_\Gamma \otimes 1 : RK_i^\Gamma(\underline{E}\Gamma) \otimes_{\mathbf{Z}} \mathbf{C} \to \bigoplus_{n=0}^{\infty} H_{i+2n}(\Gamma, F\Gamma)$$

is an isomorphism.

The origin of these notes

During the Fall 1998, I was invited to give a "Nachdiplomvorlesung" on the Baum-Connes conjecture at ETH Zürich during the Spring term 1999. At this time, the material to be covered was kind of obvious: indeed, in August 1998 came V. Lafforgue's astonishing announcement (see [59], [57]), of the proof of Conjecture 1 for some infinite groups with Kazhdan's property (T), a problem that baffled experts for more than 15 years. So the original goal of this advanced course was to give an introduction to the Baum-Connes conjecture, that would lead to Lafforgue's results. It is clear at least to me that this goal is not achieved: indeed, the 6 pages or so of Chapter 10 do not really do justice to Lafforgue's work (a nice introduction to his results, for non-experts, can be found in [80]). The reason is that I realized quickly, once the course had begun, that I had somewhat underestimated the complexity and technicality of the subject. Because of that, at least myself learned a great deal during the course.

I also fear that the written version of the notes does not really convey the flavour of the oral presentation. To be precise, during the course the 10 chapters of this book were presented in the following order: 1, 2, 4, 3, 7, 6, 8, 5, 9, 10. Indeed, I chose the *a priori* point of view of delaying Kasparov's KK-theory until the end of the course, just to contrast it with Lafforgue's Banach KK-theory. This worked basically, with the help of some hand-waving at a number of crucial points (like when I had to define the Baum-Connes assembly map). But when I started thinking about a more permanent version of the notes, it soon became clear that this was not acceptable for a text written "for eternity": a number of standards of rigour and precision are supposed to be met, that were simply absent from the hand-waved version. As a result, and with some remorse, I put Kasparov's theory in the centre of the book, as a unifying framework for K-theory and K-homology, and as a prerequisite for the rigorous definition of the analytical assembly map.

All in all, I made a sincere effort to try to make the Baum-Connes conjecture accessible to non-experts, and I hope that at least I was able to convey part of the beauty of the subject, that blends algebra, functional analysis, algebraic topology, and geometry: a subject that gives me a feeling of the unity of mathematics.

On the appendix

At some point during the course, I realized that unofficial notes were circulating, signed by a "G.M. Anonymous". That was rather transparent: the author could only be Guido Mislin, one of the most active participants in the lectures. Browsing through these notes, I realized that they contained some very interesting comments, from a topologist's point of view, on the left hand side of the Baum-Connes conjecture; for example, a comparison between various models for the universal space for proper actions (it occurred to me on this occasion that analysts and topologists do not have in mind the same model), generalities on $\Gamma - CW$-complexes, generalities on spectra in homotopy theory, and how these can be used to define the Chern character in K-homology. It was clear to me that these notes should be appended to mine, as an "output" generated by the lectures, and Guido Mislin kindly gave me permission, for which I thank him heartily.

Acknowledgements

Indira Chatterji deserves more than her share of thanks. Not only she took notes during the lectures and typed them in TeX, but she kept questioning the material, asking over and over again what she called "stupid questions" (which of course weren't). Doing so, she helped me clarify my own ideas, and obliged me to fix several points that I had pushed under the rug during the lectures. When I realized that I had to re-organize the material, she made several excellent suggestions. Actually, she convinced me that K-theory deserved more than the 3 pages of the first version, and therefore she is the author of almost all of the present Chapter 3. In a word, she did an amazing job – and, at least, she benefited from the course.

Apart from being the author of the Appendix, Guido Mislin rescued me a number of times when, cornered during the lectures, I had to admit some gap in my algebraic topology. It was always instructive to have his topologist's point of view on concepts pertaining to the left hand side of the conjecture, and I learned a lot from discussions with him during coffee breaks.

Finally, I wish to thank Marc Burger who had the idea of this set of lectures, Michael Struwe for organizing this "Nachdiplomvorlesung", and Alain-Sol Sznitman for the hospitality and excellent working conditions at the Forschungsinstitut für Mathematik (FIM).

Chapter 1

A Biased Motivation: Idempotents in Group Algebras

Let us start with a countable group Γ. We linearize Γ by associating to it the *complex group algebra* $\mathbf{C}\Gamma$, where $\mathbf{C}\Gamma$ is the $\mathbf{C}$-vector space with basis Γ. It can also be viewed as the space of functions $f : \Gamma \to \mathbf{C}$ with finite support. The product in $\mathbf{C}\Gamma$ is induced by the multiplication in Γ. Namely, for $f = \sum_{s\in\Gamma} f_s s$ and $g = \sum_{t\in\Gamma} g_t t$ elements in $\mathbf{C}\Gamma$, then

$$f * g = \sum_{s,t\in\Gamma} f_s g_t st,$$

which is the usual convolution of f and g, and thus

$$f * g(t) = \sum_{s\in\Gamma} f(s) g(s^{-1}t)$$

for all $t \in \Gamma$.

Now suppose that there is a $\gamma \in \Gamma$ which has finite order $n > 1$. Then, for w an n-th root of the unity ($w^n = 1$ in $\mathbf{C}$), the element

$$p_w = \frac{1}{n}\sum_{i=0}^{n-1} w^i \gamma^i$$

is an idempotent, that is to say satisfies $p_w^2 = p_w$. Moreover, $\sum_w p_w = 1$ and $p_w p_\sigma = 0$ for $w \neq \sigma$ and therefore, one has a ring isomorphism

$$\mathbf{C}(\mathbf{Z}/n\mathbf{Z}) \simeq \underbrace{\mathbf{C} \oplus \cdots \oplus \mathbf{C}}_{n}.$$

In particular the invertibles of $\mathbf{C}(\mathbf{Z}/n\mathbf{Z})$ are written

$$\mathbf{C}(\mathbf{Z}/n\mathbf{Z})^\times \simeq \underbrace{\mathbf{C}^\times \oplus \cdots \oplus \mathbf{C}^\times}_{n},$$

which means that there are a lot of invertibles as soon as there is torsion in Γ.

In case where Γ is torsion free, it is not known how to construct non trivial idempotents (other than 0 and 1) or non trivial invertibles (other than $\lambda\gamma$, where $\lambda \in \mathbf{C}^\times$ and $\gamma \in \Gamma$). This was turned into a conjecture:

Conjecture 2. *If Γ is torsion free, then:*

(1) *Every idempotent in $\mathbf{C}\Gamma$ is trivial.*

(2) *Every zero-divisor in $\mathbf{C}\Gamma$ is trivial.*

(3) *Every invertible in $\mathbf{C}\Gamma$ is trivial.*

Remark 1.1. (2) is stronger than (1) since an idempotent p gives a zero-divisor:

$$p(1-p) = p - p^2 = p - p = 0$$

and (3) is stronger than (2) since D. Passman (see [72]) showed that in a torsion free group, a non-trivial zero-divisor gives a non-trivial zero-root, i.e. an element $x \neq 0$ such that $x^2 = 0$, and thus $1 - x$ is a non-trivial invertible:

$$(1-x)(1+x) = 1.$$

Proposition 1.2 (H. Bass [6]). *Let A be a unital algebra over an algebraically closed field F. The following are equivalent:*

(1) *A has no idempotent except 0 and 1.*

(2) *Every finite dimensional subalgebra of A is of the form $N + F \cdot 1$, where N is a nilpotent ideal.*

(3) *Every element of A, algebraic over F, is of the form $\lambda + \nu$, where λ is a scalar and ν a nilpotent.*

(4) *Every invertible element of A, of finite order (not divisible by char(F)) is a scalar.*

Here are the known results about Conjecture 2 :

(a) Part (1) holds for the following groups (assuming they are torsion free): linear groups in characteristic 0 (H. Bass [6]); hyperbolic and CAT(0) groups; groups with cohomological dimension at most 2 over $\mathbf{Q}$ (B. Eckmann [23] and [24], Z. Marciniak [62]).

(b) Part (1), (2) and (3) hold for virtually polycyclic groups (D. Farkas and R. Snider [26]).

(c) Part (1) and (2) hold for orderable groups (D. Passman [72]).

(d) (T. Delzant [19]) Let Γ be a group of isometries of a δ-hyperbolic metric space, such that every element in $\Gamma - \{1\}$ moves points by a distance greater than 4δ. Then (3) (and thus (1) and (2)) holds for such a group Γ.

The following result is (up to now) inaccessible by algebraic tools:

Theorem 1.3 (N. Higson, G. Kasparov [34]). *If a group Γ is torsion free and a-T-menable, then $\mathbf{C}\Gamma$ has no non-trivial idempotents.*

Definition 1.4. a) An isometric action of a group Γ on a metric space X is *metrically proper* if, whenever a sequence $(\gamma_n)_{n\geq 1}$ in Γ ultimately leaves finite subsets of Γ, then the sequence $(\gamma_n x)_{n\geq 1}$ in X ultimately leaves bounded subsets of X, for every $x \in X$.

b) (M. Gromov, [28]) A group is *a-T-menable* if it admits a metrically proper, isometric action on some affine Hilbert space.

Example 1.5. For finite dimensional affine Hilbert spaces, we get crystallographic groups. By allowing infinite dimensional affine Hilbert spaces, we get a rich class of groups, containing amenable groups, free groups, discrete subgroups of $SO(n,1)$ and $SU(n,1)$, Coxeter groups, groups acting properly on trees, products of trees or simply connected CAT(0) cubical complexes (see [16]).

The terminology "a-T-menable" means that these are groups far from being groups with Kazhdan's property (T). We recall that a group Γ has *property (T)* if whenever Γ acts isometrically on some affine Hilbert space, it has a fixed point. A lattice Γ in a simple Lie group with rank greater or equal to 2 has property (T), for example $SL_n(\mathbf{Z})$ for $n \geq 3$ (see [32]).

Actually, Theorem 1.3 holds for a bigger algebra than $\mathbf{C}\Gamma$, namely the reduced C*-algebra of Γ.

Definition 1.6. For a group Γ denote by $\ell^2\Gamma$ the Hilbert space of square summable functions on Γ. The *left regular representation*, given by

$$(\lambda_\Gamma(\gamma)\xi)(s) = \xi(\gamma^{-1}s)$$

for $\gamma, s \in \Gamma$ and $\xi \in \ell^2\Gamma$ can be extended to $\mathbf{C}\Gamma$ by

$$\lambda_\Gamma(f) = \sum_{\gamma\in\Gamma} f(\gamma)\lambda_\Gamma(\gamma),$$

so that $\lambda_\Gamma(f)\xi = f * \xi$ for all $f \in \mathbf{C}\Gamma$. The representation λ_Γ of $\mathbf{C}\Gamma$ on $\ell^2\Gamma$ is by bounded operators, and faithful since $\lambda_\Gamma(f)\delta_e = f * \delta_e = f$.

The *reduced C^*-algebra* of Γ, denoted by $C_r^*\Gamma$ is the norm closure of $\lambda_\Gamma(\mathbf{C}\Gamma)$ in $\ell^2\Gamma$, namely

$$C_r^*\Gamma = \overline{\lambda_\Gamma(\mathbf{C}\Gamma)}^{\|\cdot\|_{op}},$$

where $\|\cdot\|_{op}$ denotes the operator norm on $\mathcal{B}(\ell^2\Gamma)$, given by

$$\|f\|_{op} = \sup_{\|\xi\|_2=1} \|f * \xi\|_2,$$

for all $f \in \mathbf{C}\Gamma$.

Around 1949, I. Kaplansky and R. Kadison suggested that there should be an idempotent conjecture for $C_r^*\Gamma$.

Conjecture 3 (I. Kaplansky, R. Kadison). *If Γ is torsion free, then $C_r^*\Gamma$ has no idempotents except* 0 *and* 1.

This was daring because there is no analogue, in the C*-algebraic framework, of points (2) and (3) of Conjecture 2: indeed, in a C*-algebra non isomorphic to the field of complex numbers, there are always non trivial zero-divisors and invertibles.

Example 1.7. Let Γ be a discrete abelian group, and consider its Pontryagin dual

$$\hat{\Gamma} = \mathrm{Hom}(\Gamma, S^1),$$

where $S^1 = \{z \in \mathbf{C} | \, |z| = 1\}$. With the topology induced by $(S^1)^\Gamma$, $\hat{\Gamma}$ is a compact abelian group. Denote by $C(\hat{\Gamma})$ the algebra of continuous functions on $\hat{\Gamma}$ endowed with the supremum norm. The Fourier transform is the map:

$$\begin{array}{rcl} \hat{\ }: \mathbf{C}\Gamma & \to & C(\hat{\Gamma}) \\ f & \mapsto & \hat{f} : \{\chi \mapsto \sum_{\gamma\in\Gamma} f(\gamma)\chi(\gamma)\}. \end{array}$$

In case $\Gamma = \mathbf{Z}^n$, this is the map that sends a Laurent polynomial to the associated trigonometric polynomial.

It is a classical theorem that the Fourier transform $\hat{\ }$ extends to an isometric isomorphism

$$C_r^*\Gamma \xrightarrow{\simeq} C(\hat{\Gamma}).$$

Theorem 1.8 (L. S. Pontryagin, [75]). *Let Γ be a discrete abelian group. The following are equivalent:*

(1) *The algebra $C(\hat{\Gamma})$ has no idempotent except* 0 *and* 1.

(2) *The topological space $\hat{\Gamma}$ is connected.*

(3) *The group Γ is torsion free.*

In this case Conjecture 3 corresponds to a topological statement, namely the connectedness of $\hat{\Gamma}$.

Exercise 1.9. Let A be a Banach algebra with unit. Show that the following are equivalent:

(1) There is no idempotent in A, except 0 and 1.

(2) The spectrum of every element in A is connected (recall that if $x \in A$, then $\text{spec}(x) = \{\lambda \in \mathbf{C} |\ x - \lambda \cdot 1 \text{ is not invertible in } A\}$ is a nonempty compact subset of $\mathbf{C}$).

(3) If furthermore A is a C*-algebra, this is still equivalent to saying that the spectrum of every self-adjoint element of A is an interval.

Theorem 1.10 (N. Higson, G. Kasparov, [34]). *If Γ is a torsion free, a-T-menable group, then Conjecture 3 holds.*

This theorem is deduced from the following result:

Theorem 1.11 (N. Higson, G. Kasparov, [34]). *The Baum-Connes conjecture holds for a-T-menable groups.*

Chapter 2

What is the Baum-Connes Conjecture?

2.1 A quick description

Let X be a finite simplicial complex, connected and aspherical (for each $i \geq 2$, $\pi_i(X) = 0$) and $\Gamma = \pi_1(X)$. Then X is a classifying space for Γ (or Eilenberg-Mac Lane $K(\Gamma, 1)$ space). In particular such an X is unique up to homotopy. Note that, under these assumptions Γ is torsion free.

The Baum-Connes conjecture for the group Γ states that a topological object, the K-homology of X,

$$K_i(X) = \pi_i(X_+ \wedge \mathbf{B}U),$$

[1] is isomorphic to an analytical object, the K-theory of $C_r^*\Gamma$, via a definite homomorphism

$$\mu_i^\Gamma : K_i(X) \to K_i(C_r^*\Gamma). \qquad (i = 0, 1)$$

The map μ_*^Γ is called the *index map*, the *analytical assembly map* or the *Baum-Connes assembly map*. The following chapters will be devoted to explaining the terms just used.

If Γ is the trivial group, the Baum-Connes Conjecture amounts to the K-theoretic version of the Atiyah-Singer index theorem, see [5].

It is often said that the surjectivity of μ_*^Γ has implications in analysis, while injectivity has implications in topology. Indeed, Conjecture 3 follows from the surjectivity of μ_*^Γ, as we shall see in Chapter 6. Another example where the surjectivity can be used is the following: if X is of dimension 2, then the surjectivity of

[1] where X_+ is X with a disjoint base point added, $\mathbf{B}U$ is the spectrum of topological K-theory, and π_i is the i-th homotopy group; we shall not use this homotopy-theory definition; for more about it, see the Appendix.

μ_1^Γ implies that every element in $GL_\infty(C_r^*\Gamma)$ lies in the same connected component as some diagonal matrix of the form

$$\begin{pmatrix} \gamma & 0 \\ 0 & I_\infty \end{pmatrix},$$

for some $\gamma \in \Gamma$. There is an obvious idea of a Whitehead group behind the latter statement: in algebraic K-theory, the Whitehead group $\mathrm{Wh}(\Gamma)$ is defined as

$$\mathrm{Wh}(\Gamma) = K_1^{\mathrm{alg}}(\mathbf{Z}\Gamma)/\langle \pm\gamma : \gamma \in \Gamma \rangle,$$

i.e. one looks at invertible elements modulo trivial invertibles. Therefore $\mathrm{Wh}(\Gamma) = 0$ means that every invertible is in some sense equivalent to a trivial one (compare with Conjecture 2 part (3)).

The *rational injectivity* of μ_i^Γ for $i = 0, 1$ (that is, the injectivity of the maps $\mu_i^\Gamma \otimes I$ on $K_i(X) \otimes_{\mathbf{Z}} \mathbf{Q}$) has the following consequences:

Theorem 2.1.1 (A. S. Miščenko, [66]). *If μ_i^Γ is rationally injective, then the Novikov Conjecture holds for Γ.*

Here we recall that:

Conjecture 4 (S. P. Novikov, [69]). *Higher signatures are oriented homotopy invariants, for closed oriented manifolds M with fundamental group Γ.*

Higher signatures are numbers of the form $< L \cup F^*(u), [M] >$, where L is the characteristic class appearing in the signature theorem, $f : M \to X$ is the classifying map, $u \in H^*(X, \mathbf{Q})$ and $[M]$ the fundamental class of M.

Theorem 2.1.2 (J. Rosenberg, [77]). *If μ_i^Γ is rationally injective, then the Gromov-Lawson Conjecture holds, namely: let M be a closed spin manifold; if, for some $u \in H^*(X, \mathbf{Q})$ one has*

$$< \hat{A} f^*(u), [M] > \neq 0$$

(where $\hat{A}$ is the characteristic class appearing in the index theorem for the Dirac operator on M), then M carries no metric with positive scalar curvature.

Issues on surjectivity also arose in topology, namely in the work of S. Stolz on concordance classes of metrics with positive scalar curvature (see [82]) and in the work of N. Keswani on homotopy invariance of relative eta-invariants (see [53]).

Assume that Γ is torsion free, but $B\Gamma$ is not necessarily a finite complex. Then the left hand side of the Baum-Connes Conjecture reads

$$RK_i(B\Gamma) = \varinjlim_{X \text{ compact in } B\Gamma} K_i(X). \qquad (i = 0, 1)$$

This definition is homotopy invariant, which fits with $B\Gamma$ being defined only up to homotopy.

In case where Γ has torsion, $RK_i(B\Gamma)$ and $K_i(C_r^*\Gamma)$ are in general not isomorphic, an example being provided taking $\Gamma = \mathbf{Z}/2\mathbf{Z}$, so that $C_r^*\Gamma = \mathbf{C}\Gamma = \mathbf{C}\oplus\mathbf{C}$ and $K_0(C_r^*\Gamma) \simeq \mathbf{Z}^2$. But $B\Gamma = \mathbf{P}^\infty(\mathbf{R})$, so via the Chern character in K-homology, we get an isomorphism

$$RK_0(B\Gamma) \otimes_{\mathbf{Z}} \mathbf{Q} \tilde{\rightarrow} H_{\text{even}}(\Gamma, \mathbf{Q}) = \mathbf{Q}.$$

So, to treat groups with torsion, we need a bigger group as our left hand side, namely the equivariant K-homology with Γ-compact supports of the classifying space for proper actions $\underline{E}\Gamma$.

Higson's slogan:
"$RK_*^\Gamma(\underline{E}\Gamma)$ = K-homology of $B\Gamma$ + the representation theory of finite subgroups of Γ."

Conjecture 1 (P. Baum, A. Connes, [7]). *For every countable group Γ, the analytical assembly map*

$$\mu_i^\Gamma : RK_i^\Gamma(\underline{E}\Gamma) \to K_i(C_r^*\Gamma) \qquad (i = 0, 1)$$

is an isomorphism.

2.2 Status of Conjecture 1

In the original paper of P. Baum and A. Connes [7], Conjecture 1 was proved for surface groups. Later the conjecture was proved for the following classes of groups:

(1) discrete subgroups of connected Lie groups of the form $G = RS$ (Levi-Malcev decomposition) where R is the radical and S the semi-simple part, with S locally of the form

$$S = K \times \underbrace{SO(n_1, 1) \times \cdots \times SO(n_k, 1)}_{\text{Lorentz groups}} \times \underbrace{SU(m_1, 1) \times \cdots \times SU(m_l, 1)}_{\text{Complex Lorentz groups}},$$

where K is compact (P. Julg and G. Kasparov, [43]). Special cases appear in earlier work of G. Kasparov in [45] and [49].

(2) a-T-menable groups (N. Higson and G. Kasparov, [34]).

(3) one-relator groups (C. Beguin, H. Bettaieb and A. Valette, [11]).

(4) fundamental groups of Haken 3-manifolds (this class contains all the knot groups) (independently proved by H. Oyono in [70] and J.-L. Tu in [86]).

(5) (V. Lafforgue [57]) Groups Γ satisfying the following two conditions:

- property (RD) (discussed in Chapter 8).
- Γ admits a proper, cocompact, isometric action on a strongly bolic metric space (a notion introduced by G. Kasparov and G. Skandalis [52]).

On one hand, Gromov hyperbolic groups satisfy property (RD) (a result of P. de la Harpe [31]). On the other hand, any group acting properly, cocompactly, isometrically on either a Euclidean building, or a riemannian symmetric space, satisfies the second condition. Combining those two observations, V. Lafforgue proved Conjecture 1 for "classical" hyperbolic groups, i.e. co-compact lattices in rank-one simple Lie groups $(SO(n,1), SU(n,1), Sp(n,1), F_{4(-20)})$. In particular, co-compact lattices in $Sp(n,1)$ provided the first examples of infinite property (T) groups for which Conjecture 1 holds.

Concerning higher rank lattices, it was proved by J. Ramagge, G. Robertson and T. Steger in [76] that groups acting properly, cocompactly on $\tilde{A}_2$-buildings (e. g. lattices in $SL_3(\mathbf{Q}_p)$) satisfy property (RD). These provided the first higher rank lattices for which Conjecture 1 holds. V. Lafforgue [58] then proved that cocompact lattices in $SL_3(\mathbf{R})$ and $SL_3(\mathbf{C})$ also verify property (RD), and this was extended by I. Chatterji [15] to cocompact lattices in $SL_3(\mathbf{H})$ and $E_{6(-26)}$.

Remark 2.2.1. It is not known whether the conjecture holds for $SL_n(\mathbf{Z})$ when $n \geq 3$.

Coming back to general hyperbolic groups, it was recently proved by I. Mineyev and G. Yu in [65] that every subgroup of Gromov hyperbolic group admits a proper, cocompact, isometric action on a strongly bolic metric space, hence satisfies Conjecture 1.

The rational injectivity of μ_*^Γ has been proved in many more cases, namely for groups acting properly isometrically on complete Riemannian manifolds with non-positive sectional curvature, for discrete subgroups of connected Lie groups (G. Kasparov [45]), discrete subgroups of p-adic groups (G. Kasparov and G. Skandalis [51]). N. Higson has obtained a very general criterion for the injectivity of μ_*, that we now explain.

Denote by $M^1(\Gamma)$ the set of probability measures on Γ. Since Γ is countable,

$$M^1(\Gamma) = \{p : \Gamma \to [0,1] \mid \sum_{\gamma\in\Gamma} p(\gamma) = 1\}.$$

We equip $M^1(\Gamma)$ with the weak-*-topology (since $\ell^1\Gamma = (C_0\Gamma)^*$) and also with the ℓ^1 norm.

Definition 2.2.2. A continuous action of a discrete group Γ on a compact space X is *amenable* if there exists a sequence

$$p_n : X \to M^1(\Gamma) \quad (n \geq 1)$$

of weak-*-continuous maps such that for each $\gamma \in \Gamma$ one has

$$\lim_{n\to\infty} \sup_{x\in X} \|\gamma * (p_n(x)) - p_n(\gamma \cdot x)\|_1 = 0.$$

For equivalent definitions, and for the link with Zimmer's original definition of amenable actions in the measure-theoretic setting [92], see the monograph by C. Anantharaman-Delaroche and J. Renault [1]. Note that Γ is amenable if and only if the action of Γ on the point is amenable.

Theorem 2.2.3 (N. Higson, [33]). *Assume that Γ admits an amenable action on some compact space. Then the assembly map is injective.*

There was a folk conjecture that every group Γ acts amenably on its Stone-Čech compactification $\beta\Gamma$. However a counter-example to this conjecture has been announced by M. Gromov in [29].

2.3 The Baum-Connes conjecture with coefficients

Let A be a C*-algebra on which Γ acts by automorphisms. Let $C_c(\Gamma, A)$ be the space of finitely supported functions $f : \Gamma \to A$. For $f, g \in C_c(\Gamma, A)$, $f = \sum_{s\in\Gamma} f(s)s$ and $g = \sum_{t\in\Gamma} g(t)t$, we define the *twisted convolution* by

$$f *_\alpha g = \sum_{s,t\in\Gamma} f(s)\alpha_s(g(t))st$$

where $\alpha : \Gamma \to \mathrm{Aut}(A)$. For each $t \in \Gamma$, one has that

$$(f *_\alpha g)(t) = \sum_{s\in\Gamma} f(s)\alpha_s(g(s^{-1}t)),$$

and $C_c(\Gamma, A)$ is a *-algebra whose involution is given by

$$f^*(s) = \alpha_s(f(s^{-1}))^*.$$

for all $f \in C_c(\Gamma, A)$ and $s \in \Gamma$. Similarly one can define

$$\ell^2(\Gamma, A) = \{\xi : \Gamma \to A \mid \sum_{s\in\Gamma} \xi(s)^*\xi(s) \text{ converges in } A\}.$$

The norm given by $\|\xi\| = \|\sum_{s\in\Gamma} \xi(s)^*\xi(s)\|_A$ turns $\ell^2(\Gamma, A)$ into a Banach space. The left regular representation $\lambda_{\Gamma,A}$ of $C_c(\Gamma, A)$ on $\ell^2(\Gamma, A)$ is given by

$$(\lambda_{\Gamma,A}(f)\xi)(\gamma) = \sum_{s\in\Gamma} \alpha_{\gamma^{-1}}(f(s))\xi(s^{-1}\gamma)$$

for each $f \in C_c(\Gamma, A)$ and $\xi \in \ell^2(\Gamma, A)$, so that $C_c(\Gamma, A)$ acts on $\ell^2(\Gamma, A)$ by bounded operators.

Definition 2.3.1. The *reduced crossed product* $A \rtimes_r \Gamma$ is the operator norm closure of $\lambda_{\Gamma,A}(C_c(\Gamma, A))$ in $\mathcal{B}(\ell^2(\Gamma, A))$.

Remark 2.3.2. The action of $C_c(\Gamma, A)$ on $\ell^2(\Gamma, A)$ comes from the combination of the usual action λ of Γ on $\ell^2(\Gamma, A)$, given by shifts

$$(\lambda(\gamma)\xi)(s) = \xi(\gamma^{-1}s)$$

with the action π of A on $\ell^2(\Gamma, A)$ given by

$$(\pi(a)\xi)(s) = \alpha_{s^{-1}}(a)\xi(s),$$

for $a \in A$, $\xi \in \ell^2(\Gamma, A)$ and $s, \gamma \in \Gamma$. The pair (λ, π) is called a *covariant representation* of the system $\{A, \Gamma, \alpha\}$, in the sense that it satisfies

$$\lambda(\gamma)\pi(a)\lambda(\gamma^{-1}) = \pi(\alpha_\gamma(a))$$

for all $a \in A$ and $\gamma \in \Gamma$.

Example 2.3.3. If A is commutative, then $A = C_0(Y)$, the continuous functions vanishing at infinity on a locally compact space Y. Let Γ act on Y by homeomorphisms, then $C_0(Y) \rtimes_r \Gamma$ is a suitable completion of $C_c(\Gamma \times Y)$, the space of compactly supported functions on $\Gamma \times Y$, with product given by

$$(F * G)(s, x) = \sum_{t \in \Gamma} F(t, x)G(t^{-1}s, t^{-1}x).$$

On the other hand, $\ell^2(\Gamma, C_0(Y))$ is the space of sections vanishing at infinity of the trivial field of Hilbert spaces with fiber $\ell^2\Gamma$ on Y.

In a series of papers between 1981 and 1989 [45], [47], [50], G. Kasparov defined the equivariant K-homology of $\underline{E}\Gamma$ with coefficients in A, denoted $RKK_*^\Gamma(\underline{E}\Gamma, A)$, and

$$\mu_i^{\Gamma,A} : RKK_i^\Gamma(\underline{E}\Gamma, A) \to K_i(A \rtimes_r \Gamma) \qquad (i = 0, 1)$$

which led to:

Conjecture 5 (Baum-Connes with coefficients). *For every C^*-algebra A on which a discrete group Γ acts by automorphisms, the map $\mu_i^{\Gamma,A}$ $(i = 0, 1)$ is an isomorphism.*

Conjecture 6 (Baum-Connes with commutative coefficients). *For every abelian C^*-algebra A on which a discrete group Γ acts by automorphisms, the map $\mu_i^{\Gamma,A}$ $(i = 0, 1)$ is an isomorphism.*

Remarks 2.3.4. Taking **C** for the (abelian) C*-algebra A in Conjecture 5 (or Conjecture 6) we get Conjecture 1.

Furthermore, we have the following fact, that has been stated by P. Baum, A. Connes and N. Higson in [9] and proved by H. Oyono in [70]:

Conjecture 5 and Conjecture 6 do pass to subgroups. It is not known whether Conjecture 1 is inherited by subgroups.

Conjecture 5 has been proved for classes (1) to (4) of the previous list of examples, and Conjecture 6 has been proved for "good" hyperbolic groups (e.g. co-compact lattices in rank 1 Lie groups) by V. Lafforgue in [59].

However, counter-examples to Conjecture 5 and Conjecture 6 have been announced: namely, it was proved by N. Higson, V. Lafforgue and G. Skandalis [35] that, if there exists finitely generated groups containing arbitrarily large expanders in their Cayley graph (existence of such groups is claimed by M. Gromov in [29]), then Conjecture 6 fails for such groups Γ.

2.4 Stability results on the conjecture

Theorem 2.4.1 (H. Oyono, [70]). *Suppose that a discrete group Γ acts on a tree and that Conjecture 5 holds for edges and vertex stabilizers, then it also holds for Γ.*

As a result, Conjecture 5 is stable under free or amalgamated products and HNN extensions.

Theorem 2.4.2 (H. Oyono, [71]). *Let $1 \to \Gamma_0 \to \Gamma_1 \to \Gamma_2 \to 1$ be a short exact sequence; assume that Conjecture 5 is satisfied by Γ_2 and by every subgroup H of Γ_1 containing Γ_0 as a subgroup of finite index. Then Γ_1 also satisfies Conjecture 5.*

As a result, Conjecture 5 is stable under direct products and semi-direct products when Γ_2 is torsion free.

2.5 Open questions

- Is Conjecture 6 stable under finite direct products?
- Are Conjecture 5, Conjecture 6 and Conjecture 1 stable under short exact sequences when the last group of the sequence has torsion?
- Let Γ be a discrete group with a subgroup H of finite index for which Conjecture 5, Conjecture 6 or Conjecture 1 hold, do these conjectures also hold for Γ?

Chapter 3

K-theory for (Group) C*-algebras

Part of this chapter is based on [85] and [90].

3.1 The K_0 functor

Let A be a unital algebra over $\mathbf{C}$.

Definition 3.1.1. A (right) A-module M is *projective of finite type* if there exists an A-module N and $n \geq 1$ such that

$$M \oplus N \simeq A^n$$

(as A-modules). Equivalently, there exists an idempotent $e = e^2$ in $M_n(A) = \mathrm{End}_A(A^n)$ such that $M \simeq eA^n$.

Examples 3.1.2. (1) If $A = \mathbf{C}$, a projective module of finite type is a finite dimensional vector space.

(2) If Γ is a finite group and $A = \mathbf{C}\Gamma$, then a projective module of finite type is a finite dimensional representation of the group Γ.

(3) Let X be a compact space and $A = C(X)$. By the Swan-Serre theorem (see [83]), M is a projective module of finite type over $C(X)$ if and only if $M \simeq C(X, E)$, the space of continuous sections of some complex vector bundle E over X.

Definition 3.1.3. We will write $K_0(A)$ for the *Grothendieck group of isomorphism classes of projective modules of finite type* over A, that is,

$$K_0(A) = \{(M_0, M_1)\}/\sim$$

where $(M_0, M_1) \sim (N_0, N_1)$ if there exists $n \in \mathbf{N}$ such that

$$M_0 \oplus N_1 \oplus A^n \simeq M_1 \oplus N_0 \oplus A^n.$$

Remarks 3.1.4. Recall that a *semi-group* is a set S endowed with an associative law $S \times S \to S$, we call it *abelian* if this law is commutative. For an abelian semi-group S, there exists an abelian group $G(S)$ called the *group associated to* S and a semi-group morphism $\mu : S \to G(S)$ such that for each group G and each map $\varphi : S \to G$ there is a unique homomorphism $\tilde{\varphi} : G(S) \to G$ satisfying $\tilde{\varphi} \circ \mu = \varphi$.

The group $G(S)$ can be canonically built as follows: Consider $S \times S$ with the equivalence relation $(x, y) \sim (u, v)$ if there exists an element $r \in S$ such that $x + v + r = y + u + r$ and define $G(S) = S \times S/\sim$. Then (x, x) will be the neutral element and (y, x) the inverse of (x, y). The map $\mu : S \to G(S)$ is given by $x \mapsto [(x + r, r)]$, and for a group G and a map $\varphi : S \to G$, the homomorphism $\tilde{\varphi} : G(S) \to G$ will be $\tilde{\varphi}(x, y) = \varphi(x) - \varphi(y)$.

For each $n \in \mathbf{N}$ let $P_n(A)$ be the set of idempotent matrices of $M_n(A)$, the algebra of $n \times n$ matrices over A. Let us consider $\bigcup_{n \in \mathbf{N}} P_n(A)$ with the following equivalence relation: $p \in P_n(A)$ and $q \in P_m(A)$ (for $m, n \in \mathbf{N}$) are equivalent ($p \sim q$) if one can find $k \in \mathbf{N}$, $k \geq n, m$ and $u \in GL_k(A)$ such that $p \oplus 0_{k-n} = u(q \oplus 0_{k-m})u^{-1}$ (the element $p \oplus 0_{k-n}$ is called a *trivial extension of* p, and this equivalence relation means that we require p and q to be similar up to trivial extensions). Now $\bigcup_{n \in \mathbf{N}} P_n(A)/\sim$ is an abelian semi-group with the direct sum $\oplus$ (as previously defined) as associative law, and it is equivalent to define $K_0(A)$ as the group associated with this semi-group.

Straight from the construction of $K_0(A)$ and from the previous remark about the universal group of a semi-group, by taking $G = K_0(A)$ and $\varphi = \mu$ we see that each element in $K_0(A)$ is written as a difference of two classes of idempotents $[p] - [q]$, for some $p \in P_n(A)$ and $q \in P_m(A)$. Which means that two such idempotents define the same element in $K_0(A)$ if, and only if one can find a third idempotent r such that $p \oplus r \sim q \oplus r$.

Examples 3.1.5. (1) We have that $K_0(\mathbf{C}) = \mathbf{Z}$.

(2) For Γ a finite group, $K_0(\mathbf{C}\Gamma) = \mathrm{R}(\Gamma)$, the additive group of the complex representation ring.

(3) If X is a compact topological space and $A = C(X)$, then $K_0(A) = K^0(X)$ (the topological K-theory of X, defined by means of complex vector bundles over X).

(4) Every complex vector bundle over the circle S^1 is trivial, so $K_0(C(S^1)) = K^0(S^1) \simeq \mathbf{Z}$.

(5) For $\Gamma = \mathbf{Z}^n$, $K_0(\mathbf{C}\Gamma) = \mathbf{Z}$ (here $\mathbf{C}\Gamma$ is the algebra of Laurent polynomials in n variables, which is a principal ideal domain), but $C_r^*\Gamma = C(\hat{\Gamma}) = C(\mathbf{T}^n)$ and $K_0(C_r^*\Gamma) = K^0(\mathbf{T}^n) = \mathbf{Z}^{2^{n-1}}$, see [3].

Remark 3.1.6. If A is an algebra over $\mathbf{C}$ which is not necessarily unital, it can be embedded as follows in a complex unital algebra: we consider the set $A^+ = \{(a, \lambda) | a \in A, \lambda \in \mathbf{C}\}$ with operations given as follows:

$$(a, \lambda) + (b, \mu) = (a + b, \lambda + \mu) \text{ for all } a, b \in A, \ \ \lambda, \mu \in \mathbf{C}$$
$$(a, \lambda)(b, \mu) = (ab + \mu a + \lambda b, \lambda\mu) \text{ for all } a, b \in A, \ \ \lambda, \mu \in \mathbf{C}.$$

Then the unit is $(0, 1)$. If A is an involutive algebra, we extend the involution to A^+ by:

$$(a, \lambda)^* = (a^*, \bar{\lambda}).$$

If A is a C*-algebra, then so is A^+, when the norm is given by

$$\|(a, \lambda)\| = sup\{\|xy + \lambda y\|, \|y\| = 1\},$$

which is the operator norm of A^+ acting on A.

So far we assumed and widely used the fact that the algebra A had a unit, but we can define the *K-theory group* $K_0(A)$ for any algebra A as the kernel of $\varphi_* : K_0(A^+) \to K_0(\mathbf{C}) \simeq \mathbf{Z}$, where the map φ_* is induced by $\varphi : A^+ \to \mathbf{C}$ (whose kernel is A).

If $\alpha : A \to B$ is a homomorphism of arbitrary complex algebras, we may extend it to a unital homomorphism $\alpha^+ : A^+ \to B^+$ which sends $P_n(A^+)$ to $P_n(B^+)$: it induces a homomorphism $\alpha_*^+ : K_0(A^+) \to K_0(B^+)$. Since $\varphi_B \circ \alpha^+ = \varphi_A$, the map α_*^+ induces a homomorphism $\alpha_* : K_0(A) \to K_0(B)$, so that K_0 is a covariant functor.

Definition 3.1.7. A *trace* on a $\mathbf{C}$-algebra A is a $\mathbf{C}$-linear map

$$\mathrm{Tr} : A \to \mathbf{C}$$

such that $\mathrm{Tr}(ab) = \mathrm{Tr}(ba)$ for each a, $b \in A$.

If Tr is a trace over A, we extend it to $M_n(A)$, by setting, for each $a \in M_n(A)$,

$$\mathrm{Tr}(a) = \sum_{i=1}^{n} \mathrm{Tr}(a_{ii}).$$

In other words, we consider $\mathrm{Tr} \otimes \mathrm{tr}_n$ on $A \otimes_{\mathbf{C}} M_n(\mathbf{C})$, where tr_n denotes the canonical trace on $M_n(\mathbf{C})$. If $x \in K_0(A)$, with $x = [eA^m] - [fA^n]$ we define

$$\mathrm{Tr}_*(x) = \mathrm{Tr}(e) - \mathrm{Tr}(f).$$

It is easy to see that this gives a well defined group homomorphism

$$\mathrm{Tr}_* : K_0(A) \to \mathbf{C}.$$

A trace on a C*-algebra A is *positive* if $\mathrm{Tr}(x^*x) \geq 0$ for each $x \in A$.

When A is a unital Banach algebra, there is an alternative definition of $K_0(A)$ that takes the topology into consideration. As previously, form $\bigcup_{n\in\mathbf{N}} P_n(A)$, and define a new equivalence relation $\approx$ by saying that $p \in P_n(A)$ and $q \in P_m(A)$ are in the same class for $\approx$ if they admit trivial extensions $p\oplus 0_{k-n}$ and $q\oplus 0_{k-m}$ which are in the same connected component in $P_k(A)$. It turns out that two idempotents in the same connected component of $P_n(A)$ are actually conjugate, as it follows from the following lemma together with compactness of $[0,1]$.

Lemma 3.1.8. *Let A be a unital Banach algebra and e, f be two idempotents in $M_n(A)$ such that $\|e-f\| < \dfrac{1}{\|2e-1\|}$. Then there exists an element $z \in GL_n(A)$ such that $f = z^{-1}ez$. In particular, e and f define the same class in $K_0(A)$.*

Proof. Set

$$z = \frac{1}{2}((2e-1)(2f-1)+1).$$

Then $1-z = (2e-1)(e-f)$, so that $\|1-z\| < 1$ and therefore z is invertible. Moreover, $ez = zf(=ef)$. □

Exercise 3.1.9. Show that $K_0(A)$ is the Grothendieck group associated with the semi-group $\bigcup_{n\in\mathbf{N}} P_n(A)/\approx$.

Remark 3.1.10. In a unital C*-algebra, every idempotent e is conjugate to a self-adjoint idempotent p, that is, a projector $p = p^2 = p^*$. Then, if the trace is positive, one has that $\mathrm{Tr}(e) = \mathrm{Tr}(p) = \mathrm{Tr}(p^*p) \geq 0$, and hence Tr_* maps $K_0(A)$ to $\mathbf{R}$.

Exercise 3.1.11 (The canonical trace on $C_r^*\Gamma$). On $C_r^*\Gamma$, consider the map $\tau : C_r^*\Gamma \to \mathbf{C}$ given by

$$\tau(T) = \langle T\delta_1, \delta_1\rangle .$$

Show that τ is a trace on $C_r^*\Gamma$, and that it is positive and faithful (that is, $\tau(T^*T) = 0$ if and only if $T = 0$).

3.2 The K_1 functor

Definition 3.2.1. Let A be a unital Banach algebra. We set

$$GL_\infty(A) = \varinjlim GL_n(A)$$

(with the inductive limit topology). Define, for $n \geq 1$:

$$K_n(A) = \pi_{n-1}(GL_\infty(A)),$$

where $\pi_{n-1}(GL_\infty(A))$ denotes the $(n-1)$th homotopy group of $GL_\infty(A)$, and π_0 is the group of connected components.

Example 3.2.2. (1) The group $GL_n(\mathbf{C})$ is connected. Here is a short proof that we owe to G. Valette. To connect $A \in GL_n(\mathbf{C})$ with the identity I, consider the complex affine line

$$(1-z)I + zA \qquad (z \in \mathbf{C})$$

through A and I. It meets the set of singular matrices in at most n points (since $\det((1-z)I+zA)$ is essentially the characteristic polynomial of A). So we conclude by connectedness of $\mathbf{C}$ minus n points.

From connectedness of $GL_n(\mathbf{C})$, we deduce $K_1(\mathbf{C}) = 0$.

(2) If Γ is a finite group, then $K_1(\mathbf{C}\Gamma) = 0$, since $\mathbf{C}\Gamma$ is a direct sum of matrix algebras over $\mathbf{C}$ and

$$GL_n(M_k(\mathbf{C})) = GL_{nk}(\mathbf{C})$$

is connected.

(3) Let S^d denote the d-dimensional sphere. Then $\pi_0\left(GL_n(C(S^d))\right)$ is the group of homotopy classes of continuous maps $S^d \to GL_n(\mathbf{C})$, namely $\pi_d(GL_n(\mathbf{C}))$. Since $\pi_1(GL_n(\mathbf{C})) = \mathbf{Z}$ for every $n \geq 1$, we get

$$K_1(C(S^1)) = K^1(S^1) = \mathbf{Z}$$

(a generator being given by the generator of $\pi_1(\mathbf{C}^\times)$, i.e. the map $z \mapsto z$ on S^1). Since π_2 of any connected Lie group is zero, we get $K_1(C(S^2)) = K^1(S^2) = 0$.

Remarks 3.2.3. (1) If A is a unital Banach algebra, the group $GL_1(A)$ of invertible elements is open in A. Indeed, fix $a \in GL_1(A)$ and take $b \in A$ such that $\|b-a\| < \|a^{-1}\|^{-1}$. Consider the map

$$\begin{array}{rcl} g: A & \to & A \\ z & \mapsto & a^{-1} + a^{-1}(a-b)z \end{array}$$

It is a strict contraction:

$$\|g(z) - g(z')\| \leq \|a^{-1}\| \, \|a-b\| \, \|z-z'\| < \|z-z'\|.$$

So g has a unique fixed point, which is easily seen to be the desired inverse for b.

(2) Let a be an element of A, we define the exponential of a by the absolutely convergent series

$$e^a = \sum_{n=0}^{\infty} \frac{a^n}{n!}.$$

It is an element of $GL_1(A)$, whose inverse is given by e^{-a}. If two elements a and b of A commute, then $e^{a+b} = e^a e^b$.

(3) The connected component of the unit in $GL_1(A)$ is $\langle \exp(A) \rangle$, the multiplicative subgroup of $GL_1(A)$ generated by elements of the form e^a for $a \in A$. To see it, first notice that $\langle \exp(A) \rangle$ is arc-wise connected since each element x of $\langle \exp(A) \rangle$

can be expressed as $x = e^{a_1} \dots e^{a_n}$, where $a_1 \dots a_n \in A$, and thus x is connected to the identity by the arc $x(t) = e^{ta_1} \dots e^{ta_n}$, where $t \in [0,1]$.

Now $\langle \exp(A) \rangle$ is open in $GL_1(A)$, because for an $a \in A$ satisfying $\|1-a\| < 1$ the convergent series $-\sum_{n=1}^{\infty} \frac{1}{n}(1-a)^n$ gives a logarithm for a and thus an open neighborhood V of 1 in $\exp(A)$. The multiplication by any element of $GL_1(A)$ being a homeomorphism, for each $b \in \langle \exp(A) \rangle$, Vb will be an open neighborhood of b in $\langle \exp(A) \rangle$.

Finally $\langle \exp(A) \rangle$ is closed in $GL_1(A)$ for it is an open subgroup of $GL_1(A)$.

Lemma 3.2.4. *For A a unital Banach algebra, $K_1(A)$ is an abelian group.*

Proof. First notice that the matrix $\begin{pmatrix} 1 & 0 \\ 0 & 1 \end{pmatrix}$ is connected to $\begin{pmatrix} 0 & -1 \\ 1 & 0 \end{pmatrix}$ by the arc $t \mapsto \begin{pmatrix} \cos t & -\sin t \\ \sin t & \cos t \end{pmatrix}$ in $GL_2(\mathbf{C})$, and this fact extended to the $n \times n$ matrices allows us by left (resp. right) multiplication to switch any two rows (resp. columns) of a matrix in $M_n(A)$.

Denote by $\sim$ the relation *"being in the same connected component"*. Then for $S, T \in GL_n(A)$ we have that $\begin{pmatrix} S & 0 \\ 0 & T \end{pmatrix} \sim \begin{pmatrix} T & 0 \\ 0 & S \end{pmatrix}$, so that (with I the identity $n \times n$ matrix):

$$\begin{aligned} \begin{pmatrix} ST & 0 \\ 0 & I \end{pmatrix} &= \begin{pmatrix} S & 0 \\ 0 & I \end{pmatrix}\begin{pmatrix} T & 0 \\ 0 & I \end{pmatrix} \sim \begin{pmatrix} S & 0 \\ 0 & I \end{pmatrix}\begin{pmatrix} I & 0 \\ 0 & T \end{pmatrix} \\ &= \begin{pmatrix} S & 0 \\ 0 & T \end{pmatrix} \sim \begin{pmatrix} T & 0 \\ 0 & S \end{pmatrix} = \begin{pmatrix} T & 0 \\ 0 & I \end{pmatrix}\begin{pmatrix} I & 0 \\ 0 & S \end{pmatrix} \\ &\sim \begin{pmatrix} T & 0 \\ 0 & I \end{pmatrix}\begin{pmatrix} S & 0 \\ 0 & I \end{pmatrix} = \begin{pmatrix} TS & 0 \\ 0 & I \end{pmatrix}. \end{aligned}$$

□

If $\varphi : A \to B$ is a continuous, unital homomorphism between unital Banach algebras A, B, it induces a homomorphism $\varphi_* : K_1(A) \to K_1(B)$. Again, we assumed and widely used that A had a unit, but for any Banach algebra A, the *first K-group* of A is given by the kernel of $\varphi_* : K_1(A^+) \to K_1(\mathbf{C})$, and denoted by $K_1(A)$, where φ_* is (as for K_0) induced by $\varphi : A^+ \to \mathbf{C}$. Since $K_1(\mathbf{C})$ is trivial we have $K_1(A) = K_1(A^+)$. Using this, one sees that a continuous homomorphism $\varphi : A \to B$ between arbitrary Banach algebras induces a homomorphism $\varphi_* : K_1(A) \to K_1(B)$, i.e. K_1 is a covariant functor.

3.3 Exact sequences

Proposition 3.3.1 (Weak exactness). *Let $I \subset A$ be a closed ideal. The exact sequence $0 \to I \to A \to A/I \to 0$ induces an exact sequence $K_i(I) \to K_i(A) \to K_i(A/I)$, for $i = 0, 1$.*

Proof. In order to proceed with the proof we first need to notice that for a unital C*-algebra A, if a is in the connected component of the unit, then for each surjective homomorphism $B \to A$ (where B is a Banach algebra), a is the image of an invertible element of B. Indeed, one can write $a = e^{a_1} \dots e^{a_n}$, with $b_1 \dots b_n$ pre-images of the a_i's. Then $b = e^{b_1} \dots e^{b_n}$ is invertible and pre-image of a.

To prove exactness we extend $\iota : I \to A$ and $\pi : A \to A/I$ to $\iota : I^+ \to A^+$ and $\pi : A^+ \to (A/I)^+$, so that $\pi \cdot \iota$ maps I^+ on $\mathbf{C}$, which means that the induced map $\pi_* \cdot \iota_* : K_i(I^+) \to K_i(A^+/I)$ extends the map $K_i(I^+) \to K_i(\mathbf{C})$, whose kernel is by definition $K_i(I)$. This shows that $\mathrm{im}(\iota_*) \subset \ker(\pi_*)$ and it remains to show the opposite inclusion, namely $\ker(\pi_*) \subset \mathrm{im}(\iota_*)$. We separate the cases:

$\underline{i = 0}$: Given $c \in K_0(A)$, $c = [p] - [I_n]$ with $\pi_*(c) = 0$ means that p and I_n are similar in $K_0(A/I) \subset K_0(A^+/I)$ (up to trivial extensions). The similarity can be done through an element of $\langle \exp(M_m(A^+)) \rangle$ provided that $m \in \mathbf{N}$ is big enough (for $u \in GL_k(A/I)$, $u \oplus u^{-1}$ is in the connected component of the identity in $GL_{2k}(A/I)$ and $\pi^*(p) = uqu^{-1}$ implies $\pi^*(p) \oplus 0 = (u \oplus u^{-1})(q \oplus 0)(u \oplus u^{-1})^{-1}$), so that we can lift this element $u \oplus u^{-1}$ to an element v in $GL_m(A^+)$, and $v(p \oplus 0)v^{-1}$ is $I_n \oplus 0$ modulo I, i.e. $v(p \oplus 0)v^{-1} \in M_m(I^+)$. This means that c is in the image of ι_*.

$\underline{i = 1}$: Notice that an element u in the kernel of π_* can be written as $u = a \oplus I$, where a is an element of $GL_n(A^+)$ such that $\pi(a)$ belongs to $\langle \exp(M_n(A^+/I)) \rangle$, which means that $\pi(a) = e^{a_1} \dots e^{a_m}$, with the a_i's in $M_n(A^+/I)$. Setting $b = e^{-b_m} \dots e^{-b_1} a$ where the b_i's are pre-images in $M_n(A^+)$ of the a_i's, we have that b is an element of $GL_n(A^+)$, that $[b \oplus I] = u$ (since ba^{-1} belongs to $\langle \exp(M_n(A^+)) \rangle$), and that $b \in GL_n(I^+)$ (since $\pi(b) = I$), so that $u \in \mathrm{im}(\iota_*)$. □

Proposition 3.3.2 (Long exact sequence). *Let $I \subset A$ be a closed ideal, one has the following exact sequence:*

$$K_1(I) \to K_1(A) \to K_1(A/I) \to K_0(I) \to K_0(A) \to K_0(A/I).$$

For a proof see [85] or [90]. The idea is to build a connecting homomorphism $\delta_* : K_1(A/I) \to K_0(I)$, and this is done as follows; given $a \in GL_k(A^+/I)$, we know that $a \oplus a^{-1}$ lies in $\langle \exp(M_{2k}(A^+/I)) \rangle$ and thus has a pre-image $u \in GL_{2k}(A^+)$. Let $p = u(I_k \oplus 0_k)u^{-1} \in P_{2k}(I^+)$ (see the proof of Proposition 3.3.1), we then define $\delta_*([a]) = [p] - [I_k]$.

Definition 3.3.3. The *cone* on a Banach algebra A is the set

$$CA = \{f \in C([0,1], A) \mid f(0) = 0\}.$$

It is a Banach algebra (for point-wise operations and supremum norm) and the *suspension* of A is the set

$$SA = \{f \in C([0,1], A) \mid f(0) = f(1) = 0\},$$

which is again a Banach algebra for point-wise operations and supremum norm.

Exercise 3.3.4. A Banach algebra A is *contractible* if there exists a path in $\mathrm{End}(A)$ connecting the zero map to the identity map. Notice that $\mathbf{C}$ is not contractible as a Banach algebra.

(a) Prove that SA (although in general not contractible) will be contractible if A is contractible.

(b) Prove that the cone CA is always contractible. More generally, the Banach algebra of continuous functions on a compact topologically contractible space X with target A, vanishing at one point is always contractible.

(c) Prove that $K_1(CA) = K_0(CA) = 0$.

Hint: The contractibility of CA allows to deform any matrix of $\bigcup_n P_n(CA^+)$ and $\bigcup_n GL_n(CA^+)$ into a scalar matrix.

Proposition 3.3.5. *There is a natural isomorphism between $K_1(A)$ and $K_0(SA)$.*

Proof. The C*-algebra SA being a closed ideal of CA and the map $f \mapsto f(1)$ being a surjective homomorphism from $CA \to A$ whose kernel is SA, that enables us to identify A with CA/SA. Using the long exact sequence we have that

$$K_1(SA) \to K_1(CA) \to K_1(A) \to K_0(SA) \to K_0(CA) \to K_0(A).$$

Since $K_1(CA) = K_0(CA) = 0$, we end up with the following exact sequence of groups:

$$0 \to K_1(A) \to K_0(SA) \to 0$$

which gives the sought isomorphism. □

Proposition 3.3.6. *There is a natural isomorphism between $K_0(A)$ and $K_1(SA)$.*

For a proof, see [85] or [90]. The natural isomorphism is given by the *Bott map* β_A which is defined as follows:

For a given idempotent $p \in M_n(A)$ define

$$\begin{aligned} f_p : [0,1] &\to GL_n(A^+) \\ t &\mapsto e^{2\pi i t p}. \end{aligned}$$

Since $e^{2\pi i t p} = I + (e^{2\pi i t} - 1)p$, we have that $f_p(0) = f_p(1) = I$ and hence $f_p \in M_n(SA^+)$. Now, $f_p(t)f_p(1-t) = e^{2\pi i p} = I$ implies that $f_p \in GL_n(SA^+)$ and thus we can set:

$$\begin{aligned} \beta_A : K_0(A) &\to K_1(SA) \\ [p] - [q] &\mapsto [f_p f_q^{-1}]. \end{aligned}$$

This result, combined with the previous one, drives us straight to the central theorem of K-theory:

Theorem 3.3.7 (Bott periodicity). *There is a natural isomorphism*

$$K_n(A) \simeq K_{n+2}(A)$$

for each $n \geq 0$.

Proof. Combining Proposition 3.3.5 and 3.3.6, we obtain that

$$K_i(A) \simeq K_i(S^2A). \qquad (i = 0, 1)$$

Since for $n \geq 1$, $\pi_n(GL_\infty(A)) = \pi_{n-1}(SGL_\infty(A)) = \pi_{n-1}(GL_\infty(SA))$, we immediately see that $K_n(SA) = K_{n+1}(A)$, and thus

$$K_n(A) \simeq K_n(S^2A) = K_{n+2}(A). \qquad (n \geq 1)$$

For $n = 0$, we simply use 3.3.6;

$$K_0(A) \simeq K_1(SA) = K_2(A),$$

which finishes the proof. □

We now reach an important corollary:

Corollary 3.3.8. *Let $I \subset A$ be a closed ideal. There is a six-term cyclic exact sequence*

$$\begin{array}{ccccc} K_0(I) & \xrightarrow{\iota_*} & K_0(A) & \xrightarrow{\pi_*} & K_0(A/I) \\ \delta_* \uparrow & & & & \downarrow \mu_* \\ K_1(A/I) & \xleftarrow[\pi_*]{} & K_1(A) & \xleftarrow[\iota_*]{} & K_1(I) \end{array}$$

Here the map μ_* is obtained by combining the inverse isomorphism $K_1(I) \to K_0(SI)$ (see Proposition 3.3.5) with the map

$$\delta_* \circ \beta_{A/I} : K_0(A/I) \to K_0(SI),$$

where $\beta_{A/I}$ is the Bott map for A/I, and δ_* is the connecting homomorphism for the long exact sequence. Notice that we used the fact that $S(A/I) = SA/SI$.

Example 3.3.9. To illustrate the power of six-terms exact sequences, let us give a K-theoretic proof of Jordan's simple curve theorem, suggested by A. Connes.

First step: Let Σ_g be a closed Riemann surface of genus g, and let D be a finite set of points on Σ_g. Then, for the punctured surface $\Sigma_g \backslash D$, one has that $K^0(\Sigma_g \backslash D) = \mathbf{Z}$. To see that, consider the short exact sequence

$$0 \longrightarrow C_0(\Sigma_g \setminus D) \longrightarrow C(\Sigma_g) \longrightarrow C(D) \longrightarrow 0.$$

In K-theory, it gives

$$\begin{array}{ccccc} K^0(\Sigma_g \setminus D) & \longrightarrow & K^0(\Sigma_g) & \longrightarrow & K^0(D) \\ \uparrow & & & & \downarrow \\ K^1(D) & \longleftarrow & K^1(\Sigma_g) & \longleftarrow & K^1(\Sigma_g \setminus D) \end{array}$$

Now, $K^1(D) = 0$ and $K^0(D) = \mathbf{Z}^d$. By connectedness of Σ_g, restricting a vector bundle on Σ_g to D provides a vector bundle of constant rank on D; so the image of $K^0(\Sigma_g) \to K^0(D)$ is the diagonal subgroup in $\mathbf{Z}^d$ and thus we get a short exact sequence

$$0 \longrightarrow K^0(\Sigma_g \setminus D) \longrightarrow K^0(\Sigma_g) \longrightarrow \mathbf{Z} \longrightarrow 0.$$

Now $K^0(\Sigma_g) \simeq \mathbf{Z}^2$, with generators the trivial bundle of rank 1 over Σ_g, and the bundle $p^*(L)$, where $p : \Sigma_g \to P^1(\mathbf{C}) \simeq S^2$ is a map of degree 1, and L is the canonical bundle on $P^1(\mathbf{C})$. Since both generators map to 1 in $\mathbf{Z}$, we see that $K^0(\Sigma_g \setminus D) \simeq \mathbf{Z}$.

Second step: Let U be an open, connected, orientable surface. Then $K^0(U) = \mathbf{Z}$. The following proof was supplied by G. Skandalis.

Let $f : U \to \mathbf{R}^+$ be a Morse function (i.e. a smooth function such that any critical point is non-degenerate, and which is injective on its set of critical points). Set $U_t = f^{-1}[0, t[$. Since $U = \bigcup_{t>0} U_t$, it is enough to show that, for regular values t of f, one has $K^0(U_t) = \mathbf{Z}$. But U_t is a punctured Riemann surface, so one concludes by the first step.

Third step: (Jordan's simple curve theorem). Let J be a Jordan curve in S^2, i.e. a homeomorphic image of S^1. We must show that $S^2 \setminus J$ has two connected components. It follows from the second step that, if M is an open, orientable surface, then $K^0(M) = \mathbf{Z}^c$, where c is the number of connected components of M. Since $S^2 \setminus J$ is an open, orientable surface, we must show that $K^0(S^2 \setminus J) = \mathbf{Z}^2$.

Consider for that the short exact sequence

$$0 \longrightarrow C_0(S^2 \setminus J) \longrightarrow C(S^2) \longrightarrow C(J) \longrightarrow 0$$

giving, in K-theory,

$$\begin{array}{ccccc} K^0(S^2 \setminus J) & \longrightarrow & K^0(S^2) & \longrightarrow & K^0(J) \\ \uparrow & & & & \downarrow \\ K^1(J) & \longleftarrow & K^1(S^2) & \longleftarrow & K^1(S^2 \setminus J) \end{array}$$

Now J is homeomorphic to S^1, so $K^0(J) = \mathbf{Z}$, and $K^1(J) = \mathbf{Z}$. Since any complex vector bundle over S^1 is trivial, the map $K^0(S^2) \to K^0(J)$ is onto; on the other hand $K^0(S^2) = \mathbf{Z}^2$ and $K^1(S^2) = 0$, so the above sequence unfolds as

$$0 \longrightarrow \mathbf{Z} \longrightarrow K^0(S^2 \setminus J) \longrightarrow K^0(S^2) \longrightarrow \mathbf{Z} \longrightarrow 0.$$

Working with reduced K-theory

$$\widetilde{K}^0(S^2) = \mathrm{Ker}(K^0(S^2) \to K^0(\mathrm{pt}) = \mathbf{Z}),$$

the above sequence shortens as

$$0 \longrightarrow \mathbf{Z} \longrightarrow K^0(S^2 \setminus J) \longrightarrow \widetilde{K}^0(S^2) \longrightarrow 0.$$

By Bott periodicity, $\widetilde{K}^0(S^2) = \mathbf{Z}$, so that $K^0(S^2 \setminus J) = \mathbf{Z}^2$ as desired.

Chapter 4

The Classifying Space for Proper Actions, and its Equivariant K-homology

4.1 Classifying spaces for proper actions

Let X be a Hausdorff space on which Γ acts by homeomorphisms.

Definition 4.1.1. The action of Γ on X is *proper* if for every $x, y \in X$ there exists neighborhoods U_x and U_y of x and y respectively such that the set

$$\{\gamma \in \Gamma \mid \gamma \cdot U_x \cap U_y \neq \emptyset\} \quad \text{is finite.}$$

Remark 4.1.2. Since Γ is discrete, the previous definition is equivalent to requiring the action to have a Hausdorff quotient and to satisfy the following condition: for each $x \in X$, one can find a triple (U, H, ρ), where U is a Γ-invariant neighborhood of x, H is a finite subgroup of Γ and $\rho : U \to \Gamma/H$ is a Γ-equivariant map (see e.g. p. 25 of [54]).

Example 4.1.3. (1) If $p : X \to Y$ is a (locally trivial) covering space with group Γ, then the Γ-action on X is proper (and free).

(2) If Γ is a finite group, then every action is proper.

(3) If Γ acts simplicially on a simplicial complex X, then the action is proper if, and only if the vertex stabilizers are finite.

Theorem 4.1.4 (see [54], p. 6). *If X is locally compact, then the action of Γ on X is proper if and only if for all compact subsets K, $L \subset X$, the set*

$$\{\gamma \in \Gamma \mid \gamma \cdot K \cap L \neq \emptyset\}$$

is finite.

Definition 4.1.5. A proper Γ-space $\underline{E}\Gamma$ is said to be *universal* if it is metrizable with $\underline{E}\Gamma/\Gamma$ paracompact and if for every proper metrizable Γ-space X with X/Γ paracompact there is a Γ-equivariant continuous map $X \to \underline{E}\Gamma$, unique up to Γ-equivariant homotopy. The space $\underline{E}\Gamma$ is unique up to Γ-homotopy.

Example 4.1.6. (1) If Γ is torsion free, every proper action is free, hence gives a covering with group Γ. In this case, we may take $E\Gamma$ for $\underline{E}\Gamma$, where $E\Gamma$ is the universal covering of the classifying space $B\Gamma$, using a suitable metric topology on $B\Gamma$ (see [37]).

(2) If Γ is finite, then $\underline{E}\Gamma = \{pt\}$.

(3) The space $\underline{E}\Gamma$ is Γ-equivariantly homotopic to the space $E_{\text{fin}}(\Gamma)$ introduced by P. Kropholler and G. Mislin in [56].

A criterion to identify universal spaces for proper actions is provided by the following result:

Proposition 4.1.7 (P. Baum, A. Connes, N. Higson, [9]). *A metrizable proper Γ-space X with X/Γ paracompact is universal if and only if the two following conditions hold:*

(a) *For every finite subgroup H of Γ there is an $x \in X$ stabilized by H (that is, $Hx = x$).*

(b) *The two projection maps $p_1, p_2 : X \times X \to X$ are Γ-equivariantly homotopic.*

It is easy to see that these conditions are necessary: if H is a finite subgroup of Γ, then Γ/H is a proper Γ-space, so it maps Γ-equivariantly to $\underline{E}\Gamma$; on the other hand, $\underline{E}\Gamma \times \underline{E}\Gamma$ is a proper Γ-space and the two projections $\underline{E}\Gamma \times \underline{E}\Gamma \to \underline{E}\Gamma$ are Γ-equivariant, so they must be homotopic.

Example 4.1.8. (1) If Γ acts properly on a tree X, then $\underline{E}\Gamma = |X|$ (where $|X|$ is the geometric realization of the tree with its natural metric topology). Indeed, on a tree, one has a notion of convexity, coming from the unique geodesic path between any two points: if $x, y \in |X|$, then for each $t \in [0, 1]$ the point

$$(1-t)x + ty$$

is uniquely defined. This yields a notion of barycenter, so that condition (a) of the previous proposition holds. For condition (b), the map

$$p_t(x, y) = (1-t)x + ty$$

gives a Γ-equivariant homotopy between the two projections.

(2) The previous example can be generalized to several geometric situations in which one has a notion of barycenter and the uniqueness of geodesics:

- a Riemannian symmetric space (of non-compact type).
- a Euclidean (Bruhat-Tits) building.

So if Γ acts properly isometrically on such a space X, one can take $\underline{E}\Gamma = X$.

Suppose now that X is an affine Hilbert space, separable, on which Γ acts properly isometrically. Since here also we have a notion of barycenter together with uniqueness of geodesics, we may be tempted to take $\underline{E}\Gamma = X$. However, this does not quite work, because of lack of paracompactness of X/Γ. But this can be fixed using a trick due to G. Skandalis. Let X_ω denote the space X endowed with the weak topology. Consider the embedding

$$\begin{array}{rccc} \alpha: & X \times \mathbf{R}^+ & \to & X_\omega \times \mathbf{R}^+ \\ & (x,t) & \mapsto & (x, \sqrt{\|x\|^2 + t^2}); \end{array}$$

then the inverse image of the product topology on $X_\omega \times \mathbf{R}^+$ under α is a locally compact topology on $X \times \mathbf{R}^+$. It is enough to see that:

$$\mathrm{Im}(\alpha) = \{(x,s) \in X_\omega \times \mathbf{R}^+ : \|x\| \leq s\},$$

is a locally compact subset of $X_\omega \times \mathbf{R}^+$. Take $y = (x,t) \in X \times \mathbf{R}^+$, a closed fundamental neighborhood of $\alpha(y)$ in $X_\omega \times \mathbf{R}^+$ is of the form:

$$V_{(x_1,\ldots,x_n,\epsilon,\delta)} = \{(x',s') \in X_\omega \times \mathbf{R}^+,\ |s-s'| \leq \epsilon, |\langle x-x', x_i\rangle| \leq \delta,\ 1 \leq i \leq n\}$$

where $x_1, \ldots, x_n \in X$ and $\epsilon, \delta \in \mathbf{R}^+$, so that

$$U_{\alpha(y)} = V_{(x_1,\ldots,x_n,\epsilon,\delta)} \cap \mathrm{Im}(\alpha)$$

is a closed neighborhood of $\alpha(y)$ in $\mathrm{Im}(\alpha)$, and $U_{\alpha(y)}$ is compact since it is contained in a closed ball of radius $s+\epsilon$ (which is weakly compact).

Let us now show that Γ acts by homeomorphisms on $X \times \mathbf{R}^+$ endowed with this locally compact topology. The linear part of the action being anyway by homeomorphisms, it remains to check that the affine part is by homeomorphisms as well, which is clear since for $a \in X$, the transported translation t_a on $\mathrm{Im}(\alpha)$ reads

$$\tilde{t}_a(x,s) = \alpha \circ t_a \circ \alpha^{-1}(x,s) = (x+a, \sqrt{\|a\|^2 + 2\mathrm{Re}\,\langle a,x\rangle + s^2}).$$

Finally, this action is proper since a compact subset in $X_\omega \times \mathbf{R}^+$ is bounded in $X \times \mathbf{R}^+$. So, if Γ acts properly isometrically on X, a model for $\underline{E}\Gamma = X$ is $X \times \mathbf{R}^+$ with this new topology (this has been used by N. Higson and G. Kasparov [34] in the proof of Conjecture 1 for a-T-menable groups).

(3) If the group Γ is Gromov hyperbolic, one may take the Rips complex as a model for $\underline{E}\Gamma$. Here we recall that the Rips complex is given as follows. Take $S \subset \Gamma$ a finite generating set. This gives a metric on Γ, namely the word length metric associated to the generating set S, which is given by

$$\ell_S(\gamma) = \min\{n \in \mathbf{N} \mid \gamma = s_1 \ldots s_n \text{ and } s_1, \ldots, s_n \in S \cup S^{-1}\}.$$

Now fix $R \in \mathbf{R}^+$, a $(k+1)$-tuple $(\gamma_0, \ldots, \gamma_k) \in \Gamma^{k+1}$ will be a k-simplex if

$$\ell_S({\gamma_i}^{-1}\gamma_j) \leq R$$

for each $i, j = 0, \ldots, k$. This defines a proper metric Γ-space $M(R)$. When Γ is hyperbolic, for R big enough one may take $M(R) = \underline{E}\Gamma$.

(4) For arbitrary countable groups Γ, one may take

$$\underline{E}\Gamma = \{f : \Gamma \to [0,1] \text{ with finite support such that } \sum_{\gamma \in \Gamma} f(\gamma) = 1\}$$

(i.e. finitely supported probability measures) with metric given by

$$\|f - g\|_\infty = \sup_{\gamma \in \Gamma} |f(\gamma) - g(\gamma)|,$$

for all $f, g \in \underline{E}\Gamma$. This is Γ-equivariantly homotopic to the geometric realization of the simplicial complex whose k-simplices are given by the subsets of Γ of cardinality $(k+1)$, see Theorem A.2.1.

Now we will give the proof of the universality of this latter model.

We first check that the action of Γ on $\underline{E}\Gamma$ is proper. Fix $f \in \underline{E}\Gamma$ and denote by Γ_f its stabilizer, which is a finite subgroup of Γ. Set

$$R = \inf\{\|f - \gamma \cdot f\|_\infty \mid \gamma \in \Gamma - \Gamma_f\};$$

clearly $R > 0$. For $\epsilon > 0$, define

$$U_\epsilon = \{g \in \underline{E}\Gamma \text{ such that there exists } \gamma \in \Gamma \text{ with } \|g - \gamma \cdot f\|_\infty < \epsilon\};$$

it is an open, Γ-invariant subset of $\underline{E}\Gamma$. Moreover, for $\epsilon < R/2$, the open set U_ϵ is such that, for $g \in U_\epsilon$, the element $\gamma \in \Gamma$ with $\|g - \gamma \cdot f\|_\infty < \epsilon$ is unique modulo Γ_f. Sending g to the coset $\gamma\Gamma_f$ then defines a continuous Γ-equivariant map $U_\epsilon \to \Gamma/\Gamma_f$. So, by Remark 4.1.2, $\underline{E}\Gamma$ is a proper Γ-space.

Let X be a proper Γ-space. We have to show that there exists a continuous Γ-equivariant map $X \to \underline{E}\Gamma$, which is unique up to Γ-equivariant homotopy. Uniqueness is clear, since $\underline{E}\Gamma$ is a convex set on which Γ acts affinely. For the existence, denote by W the disjoint union of the Γ/H's, where H runs along finite subgroups of Γ. Define a Γ-equivariant map

$$\phi : W \to \underline{E}\Gamma$$

by sending the coset γH to the uniform probability measure on γH. Since X/Γ is paracompact, there exists a countable partition of unity $(\alpha_k)_{k \geq 1}$ on X, consisting of Γ-invariant functions and such that, for every $k \geq 1$, there is a Γ-equivariant continuous map $\psi_k : \alpha_k^{-1}]0,1] \to W$. Then the map

$$\begin{array}{rccc} \Psi : & X & \to & \underline{E}\Gamma \\ & x & \mapsto & \sum_{k=1}^{\infty} \alpha_k(x)(\phi \circ \psi_k)(x) \end{array}$$

is continuous and Γ-equivariant. This proof is slightly more direct than the one in paragraph 2 of [9].

4.2 Equivariant K-homology

Let X be a locally compact space, endowed with a proper action of Γ. Let $C_0(X)$ be the algebra of continuous functions on X, vanishing at infinity. The goal is to define Γ-equivariant K-homology groups $K_i^\Gamma(X)$ (where $i = 0, 1$). Cycles were defined by M. Atiyah in [4]. The equivalence relation on these cycles turning them into an abelian group in duality with equivariant K-theory was worked out by G. Kasparov in [45]. A wealth of informations on analytic K-homology can be found in the recent book by N. Higson and J. Roe [36].

Definition 4.2.1. A *generalized elliptic* Γ*-operator* over X is a triple (U, π, F) where:

- U is a unitary representation of Γ on some Hilbert space $\mathcal{H}$.
- π is a *-representation of $C_0(X)$ by bounded operators on $\mathcal{H}$ (that is, $\pi(\overline{f}) = \pi(f)^*$) which is covariant, in the sense that

$$\pi(f \circ \gamma^{-1}) = U_\gamma \pi(f) U_\gamma^{-1}$$

  for all $f \in C_0(X)$ and $\gamma \in \Gamma$.
- F is a bounded, self-adjoint operator on $\mathcal{H}$, which is Γ-equivariant (that is, $FU_\gamma = U_\gamma F$ for each $\gamma \in \Gamma$) and such that the operators

$$\pi(f)(F^2 - 1) \text{ and } [\pi(f), F]$$

  are compact for all $f \in C_0(X)$.

Such a cycle (U, π, F) is *even* if the Hilbert space $\mathcal{H}$ is $\mathbf{Z}_2$-graded, and U, π preserve the graduation whereas F reverses it. This means that $\mathcal{H} = \mathcal{H}_0 \oplus \mathcal{H}_1$ and, in that decomposition:

$$U = \begin{pmatrix} U_0 & 0 \\ 0 & U_1 \end{pmatrix}, \quad \pi = \begin{pmatrix} \pi_0 & 0 \\ 0 & \pi_1 \end{pmatrix}, \quad F = \begin{pmatrix} 0 & P^* \\ P & 0 \end{pmatrix}.$$

A cycle will be *odd* otherwise. Even cycles will be used to build $K_0^\Gamma(X)$, and odd cycles to build $K_1^\Gamma(X)$.

Remarks 4.2.2. If X is compact (that is, Γ is finite), conditions in the definition of a cycle can be replaced by $(F^2 - 1)$ and $[\pi(f), F]$ compact. This means that F is invertible modulo compact operators, i.e. F is a *Fredholm operator*.

A word about functoriality: if $h : X \to Y$ is a proper Γ-equivariant and continuous map between locally compact Γ-spaces, then the map $h^* : C_0(Y) \to C_0(X)$ defined as $h^*(f) = f \circ h$, sends a cycle (U, π, F) over X to the cycle $(U, \pi \circ h^*, F)$ over Y. The theory is therefore covariant.

Example 4.2.3. (1) Let H be a finite subgroup of Γ, set $X = \Gamma/H$ (a discrete, proper Γ-space). Let ρ be a finite dimensional representation of H on some vector space V_ρ. Let $\mathcal{H}_0$ be the space of the induced representation,

$$\mathcal{H}_0 = \{\xi \in \ell^2(\Gamma, V_\rho) \mid \xi(\gamma \cdot h) = \rho(h^{-1})\xi(\gamma)\ \forall\, h \in H, \gamma \in \Gamma\},$$

where $\ell^2(\Gamma, V_\rho)$ denotes the space of functions

$$\xi : \Gamma \to V_\rho \quad \text{such that} \quad \sum_{x \in X} \|\xi(x)\|^2 < \infty.$$

In other terms, this space $\mathcal{H}_0$ is the space of ℓ^2-sections of the induced vector bundle $\Gamma \times_H V_\rho$ over X. Then U_0 comes from the natural left action of Γ, whereas $\pi_0(f)\xi = f \cdot \xi$ (pointwise multiplication). Now, take

$$\mathcal{H}_1 = 0\,,\ \mathcal{H} = \mathcal{H}_0 \oplus \mathcal{H}_1 = \mathcal{H}_0$$

and $F = 0$. We claim that $\beta_{H,\rho} = (U_0, \pi_0, F)$ as defined above is an even cycle. We only have to check that π_0 is a representation by compact operators, which is true since for each $f \in C_0(X)$, the operator $\pi_0(f)$ can be represented as a block diagonal matrix, with each block being scalar and given by the values of f, and these do go to zero at infinity since we assumed $f \in C_0(X)$.

(2) Let $\Gamma = \{e\}$, $X = S^1$ with Lebesgue measure of mass one, and $\mathcal{H} = L^2(S^1)$. Consider the trigonometric basis $(e^{2\pi in\theta})_{n\in\mathbf{Z}}$. For the representation π of $C(S^1)$, we take pointwise multiplication. Set

$$F(e^{2\pi in\theta}) = \begin{cases} e^{2\pi in\theta} & n > 0 \\ -e^{2\pi in\theta} & n < 0 \\ 0 & n = 0 \end{cases},$$

that is, $F = \mathrm{diag}(\mathrm{sign}(n))_{n\in\mathbf{Z}}$, and $1 - F^2$ is the projection onto the constant functions in $L^2(S^1)$. The claim is that (π, F) is an odd cycle over S^1. Set $A = \{f \in C(S^1)$ such that $[\pi(f), F]$ is compact $\}$, we have to show that $A = C(S^1)$. Since the set of compact operators is a norm-closed ideal in the algebra of bounded operators on $L^2(S^1)$, we see that A is a norm-closed *-subalgebra of $C(S^1)$. Define

$$f_1 = e^{2\pi i\theta} \in C(S^1).$$

The operator $[\pi(f_1), F]$ is of rank 2 (hence is compact), so that $f_1 \in A$ and thus A contains the *-subalgebra generated by f_1, i.e. the algebra of trigonometric polynomials; since the latter is norm-dense in $C(S^1)$ (Weierstraß' theorem), we have $A = C(S^1)$. Hence (π, F) is an odd cycle, which will turn out to be the generator of $K_1(S^1) \simeq \mathbf{Z}$. Notice that on $L^2(S^1)$ one has an unbounded operator

$$D = -i\frac{d}{d\theta},$$

so $D(e^{2\pi in\theta}) = 2\pi ne^{2\pi in\theta}$ and $F = sign(D)$ (i.e. $F = \dfrac{D}{|D|}$ on $(\mathbf{C} \cdot 1)^\perp$).

(3) Let $X = \mathbf{R}$ and $\Gamma = \mathbf{Z}$ (acting by integer translations), $\mathcal{H} = L^2(\mathbf{R})$ (with respect to Lebesgue measure) and π is the representation of $C_0(\mathbf{R})$ by pointwise multiplication. The unbounded operator

$$D = -i\frac{d}{dt}$$

becomes, under Fourier transform, the multiplication by the dual variable λ on $L^2(\hat{\mathbf{R}})$. Let G be the operator of multiplication by $\operatorname{sign}(\lambda)$ on $L^2(\hat{\mathbf{R}})$ and let F be the operator on $L^2(\mathbf{R})$ obtained by inverse Fourier transform (F is the Hilbert transform on $L^2(\mathbf{R})$). Then $F^2 = 1$ and $FU_n = U_nF$ for each $n \in \mathbf{Z}$. It remains to check that the operator $[\pi(f), F]$ is compact for each $f \in C_0(\mathbf{R})$. First start with f a Schwartz function (i.e. a function having all its derivatives decreasing faster than the inverse of any polynomial). Under Fourier transform, $\pi(f)$ goes to convolution $\rho(\hat{f})$ by the function $\hat{f}$, so that $[\pi(f), F]$ goes to the following operator on $L^2(\hat{\mathbf{R}})$:

$$[\rho(\hat{f}), G](\xi)(\lambda) = \int_{-\infty}^{\infty} (\operatorname{sign}(\lambda) - \operatorname{sign}(\mu))\hat{f}(\lambda - \mu)\xi(\mu)d\mu.$$

This is an operator with kernel given by

$$K(\lambda, \mu) = (\operatorname{sign}(\lambda) - \operatorname{sign}(\mu))\hat{f}(\lambda - \mu).$$

Now, $K(\lambda, \mu)$ is in $L^2(\mathbf{R}^2)$ since

$$\int_{\mathbf{R}^2} |K(\lambda, \mu)|^2 d\lambda \, d\mu = 2\int_0^{\infty} \int_{-\infty}^{0} 4|\hat{f}(\lambda - \mu)|^2 d\lambda \, d\mu$$

and $\hat{f}$ being a Schwartz function, the right hand side of the last equality is smaller than

$$C + C' \int_1^{\infty} \int_{\mu}^{\infty} \lambda^{-3} d\lambda \, d\mu$$

which is finite. So we have proved that the operator is Hilbert-Schmidt, and in particular is compact. By density of the space of Schwartz functions in $C_0(\mathbf{R})$, the operators $[\pi(f), F]$ are compact, for each $f \in C_0(\mathbf{R})$. The triple (U, π, F) is an odd cycle over $\mathbf{R}$. Eventually there will be a canonical isomorphism $K_1^{\mathbf{Z}}(\mathbf{R}) \to K_1(S^1)$ sending this cycle to the one defined in example (2).

(4) Let X be a Γ-proper Riemannian manifold, Γ-compact (that is, $\Gamma\backslash X$ is compact), E a Γ-vector bundle over X, endowed with a Γ-invariant Hermitian structure and D an elliptic differential operator on $C_c^{\infty}(X, E)$, the space of smooth sections. This means that if one writes locally

$$D = \sum_{|\alpha| \leq m} a_\alpha D^\alpha$$

with the a_α local smooth sections of End(E), its principal symbol

$$a_m(x,\xi) = \sum_{|\alpha|=m} a_\alpha(x)\xi^\alpha$$

is invertible for each $x \in X$ and $\xi \in T_x^* - \{0\}$. If D is Γ-invariant and symmetric (with respect to the scalar product), then D is essentially self-adjoint on $L^2(X,E)$ and hence one can define thanks to the spectral theorem

$$F = \frac{D}{\sqrt{1+D^2}}$$

which is a pseudo-differential operator of order 0, bounded and self-adjoint. Let π be the representation of $C_0(X)$ by pointwise multiplication on $\mathcal{H} = L^2(X,E)$. Denote by $C_c^\infty(X)$ the space of smooth, compactly supported functions on X. For $f \in C_c^\infty(X)$, the operator $[\pi(f), F]$ is pseudo-differential of lower order (from the symbol exact sequence) so it defines a compact operator on $L^2(X,E)$ (indeed, one may assume that supp(f) lies in an open chart, so we may appeal to corresponding results on compact manifolds). Now,

$$F^2 - 1 = \frac{-1}{1+D^2},$$

and because D is elliptic, this operator is "locally compact" in the sense that $\pi(f)(F^2-1)$ is compact for all $f \in C_c^\infty(X)$. Hence (U,π,F) is a cycle. Geometry provides many such operators D: de Rham operators, signature operators, Dirac operators, Dolbeault operators. See [89] for the theory of elliptic operators on manifolds.

Definition 4.2.4. *(i)* A cycle $\alpha = (U,\pi,F)$ over X is *degenerate* if for each $f \in C_0(X)$ one has

$$[\pi(f), F] = 0 \text{ and } \pi(f)(F^2-1) = 0.$$

(ii) Two cycles $\alpha_0 = (U_0,\pi_0,F_0)$ and $\alpha_1 = (U_1,\pi_1,F_1)$ are said to be *homotopic* if $U_0 = U_1$, $\pi_0 = \pi_1$ and there exists a norm continuous path $(F_t)_{t\in[0,1]}$, connecting F_0 to F_1, such that for each $t \in [0,1]$ the triple $\alpha_t = (U_0,\pi_0,F_t)$ is a cycle (of the same parity).

(iii) The cycles α_0 and α_1 are said to be *equivalent* and denoted $\alpha_0 \sim \alpha_1$ if there exists two degenerate cycles β_0 and β_1 such that, up to unitary equivalence, $\alpha_0 \oplus \beta_0$ is homotopic to $\alpha_1 \oplus \beta_1$.

(iv) We will write $K_0^\Gamma(X)$ for the set of equivalence classes of even cycles over X and $K_1^\Gamma(X)$ for the set of equivalence classes of odd cycles over X. In case Γ is the trivial group, we just write $K_0(X)$ and $K_1(X)$. We shall see in Proposition 4.2.7 below that the $K_i^\Gamma(X)$'s are indeed abelian groups.

Remark 4.2.5. It is a deep result of G. Kasparov in [47] that this definition of K_0^Γ and K_1^Γ is invariant under proper Γ-homotopies, in the sense that if $f, g : X \to Y$ are proper Γ-homotopic maps, then

$$f_* = g_* : K_i^\Gamma(X) \to K_i^\Gamma(Y)$$

for $i = 0, 1$. One could give a weaker definition of homotopy, by dropping the requirement of the representations π_0 and π_1 to be equal. Instead, one requires to admit a continuous (in a suitable sense) path $t \mapsto \pi_t$ of representations, joining π_0 to π_1. In [47], G. Kasparov proves that the groups constructed by means of this weaker definition are isomorphic to the ones that we previously defined.

Example 4.2.6. Let $\alpha = (U, \pi, F)$ be an even cycle, with

$$F = \begin{pmatrix} 0 & P^* \\ P & 0 \end{pmatrix}.$$

If we view it as an odd cycle by forgetting about the $\mathbf{Z}_2$-graduation, then α is homotopic to a degenerate cycle via the path of operators

$$F_t = \begin{pmatrix} \sin(\frac{\pi t}{2}) & P^* \cos(\frac{\pi t}{2}) \\ P\cos(\frac{\pi t}{2}) & -\sin(\frac{\pi t}{2}) \end{pmatrix},$$

for $t \in [0, 1]$.

Proposition 4.2.7. *The sets $K_i^\Gamma(X)$ are abelian groups, for $i = 0, 1$.*

Proof. Addition is induced by direct sum; it is commutative because for two cycles α and β, the cycle $\alpha \oplus \beta$ is unitarily equivalent to $\beta \oplus \alpha$. Two degenerate cycles β_0 and β_1 are equivalent because $\beta_0 \oplus \beta_1$ is unitarily equivalent to $\beta_1 \oplus \beta_0$, hence $\beta_0 \sim \beta_1$. The class of degenerate cycles provides the neutral element for the addition. Now let us describe the opposites:

<u>$i = 0$</u>: Let α be an even cycle. To construct $-\alpha$, reverse the grading. If

$$F = \begin{pmatrix} 0 & P^* \\ P & 0 \end{pmatrix}, \quad \text{then} \quad \alpha \oplus -\alpha = \begin{pmatrix} 0 & \begin{pmatrix} P^* & 0 \\ 0 & P \end{pmatrix} \\ \begin{pmatrix} P & 0 \\ 0 & P^* \end{pmatrix} & 0 \end{pmatrix}$$

in the decomposition

$$\underbrace{\mathcal{H}_0 \oplus \mathcal{H}_1}_{\text{even part}} \oplus \underbrace{\mathcal{H}_1 \oplus \mathcal{H}_0}_{\text{odd part}}.$$

It is homotopic to a degenerate cycle via the path

$$F_t = \begin{pmatrix} 0 & \begin{pmatrix} P^*\cos(\frac{\pi t}{2}) & \sin(\frac{\pi t}{2}) \\ -\sin(\frac{\pi t}{2}) & P\cos(\frac{\pi t}{2}) \end{pmatrix} \\ \begin{pmatrix} P\cos(\frac{\pi t}{2}) & -\sin(\frac{\pi t}{2}) \\ \sin(\frac{\pi t}{2}) & P^*\cos(\frac{\pi t}{2}) \end{pmatrix} & 0 \end{pmatrix}.$$

$\underline{i=1}$: The opposite of a cycle $\alpha = (U, \pi, F)$ is $-\alpha = (U, \pi, -F)$. Indeed, $\alpha \oplus -\alpha$ is homotopic to a degenerate cycle via the path

$$F_t = \begin{pmatrix} F\cos(\frac{\pi t}{2}) & \sin(\frac{\pi t}{2}) \\ \sin(\frac{\pi t}{2}) & -F\cos(\frac{\pi t}{2}) \end{pmatrix},$$

for $t \in [0,1]$. □

Exercise 4.2.8. If $\alpha = (U, \pi, F)$ is an even cycle, then it is homotopic to $(U, \pi, -F)$.

Proposition 4.2.9. *If Γ acts freely on X (e. g. if Γ is torsion free), then there exists a canonical isomorphism*

$$K_i^\Gamma(X) \simeq K_i(\Gamma\backslash X). \qquad (i = 0, 1)$$

For a proof see [50].

Example 4.2.10. Assume that Γ is finite, $X = \{\mathrm{pt}\}$. We have that $K_0^\Gamma(\mathrm{pt})$ is the additive group of the representation ring $R(\Gamma)$. If

$$[U_0] - [U_1] \in R(\Gamma),$$

we associate to it the even cycle $(U_0 \oplus U_1, \lambda, 0)$, where λ denotes the scalar multiplication of $\mathbf{C}$. That gives us a map $R(\Gamma) \to K_0^\Gamma(\mathrm{pt})$. Conversely, if

$$(U, \lambda, F = \begin{pmatrix} 0 & P^* \\ P & 0 \end{pmatrix})$$

is an even cycle, we associate to it its Γ-index

$$\mathrm{Ind}_\Gamma(F) = [\ker P] - [\ker P^*]$$

viewed as an element in $R(\Gamma)$.

On the other hand, $K_1^\Gamma(\mathrm{pt}) = 0$. Indeed, an odd cycle is given by some Γ-invariant operator F such that $F^2 - 1$ is compact. The essential spectrum of F (that is, the part of the spectrum which is invariant under perturbations of compact operators), is $\{\pm 1\}$. But it is also the set of limit points or eigenvalues with infinite multiplicity in the spectrum of F. So, in the spectrum of F there are gaps between -1 and 1. Let $[a, b]$ be contained in such a gap. Consider the continuous function f defined as follows:

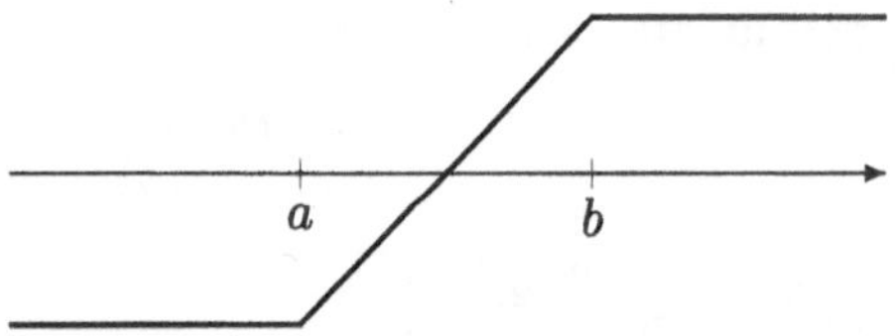

i.e. f interpolates linearly between -1 and 1 on $[a,b]$. By functional calculus, $f(F) = G$ is such that $G^2 = 1$ because $f^2 = 1$ on the spectrum of F. But $F - G$ is compact, and F is homotopic to the degenerate cycle given by G via the homotopy

$$F_t = (1-t)F + tG.$$

Definition 4.2.11. Let Y be a topological Hausdorff space on which Γ acts properly. For $i = 0, 1$, define

$$RK_i^\Gamma(Y) = \lim_{\overrightarrow{X}} K_i^\Gamma(X)$$

where X runs along the inductive system of Γ-compact subsets of Y. This is the *Γ-equivariant K-homology with compact supports*.

It is important to notice (and this contributes to the value of Conjecture 1) that $RK_i^\Gamma(\underline{E}\Gamma)$ can be computed (up to torsion) as something homological. Indeed, let $F\Gamma$ be the $\mathbf{C}$-vector space generated by elements of finite order in Γ (note that $1 \in \Gamma$ has finite order). The action of Γ by conjugation turns $F\Gamma$ into a Γ-module.

Theorem 4.2.12 (P. Baum and A. Connes [8]). *There exists an equivariant Chern character*

$$\mathrm{Ch}_*^\Gamma : RK_i^\Gamma(\underline{E}\Gamma) \to \bigoplus_{n=0}^{\infty} H_{2n+i}(\Gamma, F\Gamma)$$

which is an isomorphism after tensoring the domain with $\mathbf{C}$. Moreover, it is compatible with the Chern character in homology, in the sense that the following diagram commutes:

$$\begin{array}{ccccc}
RK_i(B\Gamma) & \xrightarrow{\simeq} & RK_i^\Gamma(E\Gamma) & \xrightarrow{\text{canonical}} & RK_i^\Gamma(\underline{E}\Gamma) \\
\downarrow{\scriptstyle \mathrm{Ch}_*} & & & & \downarrow{\scriptstyle \mathrm{Ch}_*^\Gamma} \\
\bigoplus_{n=0}^{\infty} H_{2n+i}(\Gamma, \mathbf{C}) & & \xrightarrow{\mathbf{C}\hookrightarrow F\Gamma,\ 1\mapsto 1} & & \bigoplus_{n=0}^{\infty} H_{2n+i}(\Gamma, F\Gamma)
\end{array}$$

Remarks 4.2.13. (1) It follows from Theorem 4.2.12 that the map

$$RK_i^\Gamma(E\Gamma) \to RK_i^\Gamma(\underline{E}\Gamma)$$

is rationally injective. A direct proof is given in M. Matthey's thesis, [63].

(2) After identifying $RK_i(B\Gamma)$ with something homotopical (as in the beginning of Chapter 2), we have that

$$RK_i(B\Gamma) \otimes_{\mathbf{Z}} \mathbf{C} \simeq \bigoplus_{n=0}^{\infty} H_{2n+i}(\Gamma, \mathbf{C}).$$

see also Lemma A.4.2 in the Appendix.

(3) If Γ is finite, then $\underline{E}\Gamma = \{\mathrm{pt}\}$, and this theorem reduces to the classical (but not quite obvious) fact that the number of irreducible representations of Γ is equal to the number of conjugacy classes.

Exercise 4.2.14 (A. Connes, [17]). Define a non self-adjoint cycle by dropping the condition that $F = F^*$ in Definition 4.2.1. If

$$(U, \pi = \begin{pmatrix} \pi_0 & 0 \\ 0 & \pi_1 \end{pmatrix}, F = \begin{pmatrix} 0 & Q \\ P & 0 \end{pmatrix})$$

is an even non self-adjoint cycle, then it is equivalent to another even non self-adjoint cycle say $(\tilde{U}, \tilde{\pi}, \tilde{F})$, where $\tilde{F}$ satisfies $\tilde{F}^2 = 1$.

Hint: Let $\bar{\mathcal{H}}$ be $\mathcal{H}$ with the $\mathbf{Z}_2$-grading reversed. On $\bar{\mathcal{H}} \oplus \mathcal{H}$, set

$$\tilde{\pi} = \pi \oplus 0 = \begin{pmatrix} \tilde{\pi}_0 & 0 \\ 0 & \tilde{\pi}_1 \end{pmatrix}, \quad \text{where} \quad \tilde{\pi}_i = \begin{pmatrix} \pi_i & 0 \\ 0 & 0 \end{pmatrix} \quad \text{for} \quad i = 0, 1.$$

Set

$$\tilde{F} = \begin{pmatrix} 0 & \begin{pmatrix} (2-QP)Q & 1-QP \\ 1-PQ & -P \end{pmatrix} \\ \begin{pmatrix} P & 1-PQ \\ 1-QP & (QP-2)Q \end{pmatrix} & 0 \end{pmatrix}$$

Check that $\tilde{F}^2 = 1$, that $(\tilde{U}, \tilde{\pi}, \tilde{F})$ is an even cycle and that $(\tilde{F} - (F \oplus 0))\tilde{\pi}(f)$ is compact for each $f \in C_0(X)$. Now $(1-t)\tilde{F} + t(F \oplus 0)$ gives the homotopy between $(\tilde{U}, \tilde{\pi}, \tilde{F})$ and $(U, \pi, F) \oplus 0$.

Exercise 4.2.15 (G. Skandalis, [79]). If (U, π, F) is a non self-adjoint cycle with $F^2 = 1$, show that there exists a self-adjoint cycle (U, π, F'), such that $F'^2 = 1$, and which is homotopic to (U, π, F).

Hint: Take $F' = F(FF^*)^{1/2}$. Using that $FF^* = (F^*F)^{-1}$, show that F' is unitary and that $F'^2 = 1$. Moreover, $F_t = F(FF^*)^{t/2}$ gives the homotopy, for $t \in [0, 1]$.

Exercise 4.2.16. Any cycle (U, π, F) is equivalent to (U', π', F'), a cycle in which the representation π' is essential (where *essential* means that $\pi'(C_0(X))\mathcal{H}$ is dense in $\mathcal{H}$).

Exercise 4.2.17. Let (U, π, F) be an odd cycle, show that the even cycle

$$(U \oplus U, \pi \oplus \pi, G = \begin{pmatrix} 0 & F \\ F & 0 \end{pmatrix})$$

is equivalent to a degenerate cycle.

Hint: (a) Show that α is equivalent to a cycle

$$(U' \oplus U', \pi' \oplus \pi', \begin{pmatrix} 0 & F' \\ F' & 0 \end{pmatrix})$$

where $F' = F'^*$ is invertible.

(b) Using (a) we may assume $F = F^*$ invertible and define

$$\beta = (U \oplus U, \pi \oplus \pi, \tilde{G} = \begin{pmatrix} 0 & F \\ F^{-1} & 0 \end{pmatrix}).$$

Show that $\alpha \oplus \beta$ is degenerate. Clearly $\alpha = -\alpha$ and thus β is equivalent to α.

(c) Using Exercise 4.2.15, show that β is homotopic to a degenerate cycle.

Exercise 4.2.18. Show that $K_1(S^1) = \mathbf{Z}$.

Hint: For an odd cycle (U, π, F) in $K_1(S^1)$, we may assume that $F^2 = 1$, so that $P = (F+1)/2$ is an idempotent and $[\pi(f), P]$ is compact for each $f \in C(S^1)$. Define

$$\alpha(f) = P\pi(f)|_{P\mathcal{H}} : P\mathcal{H} \to P\mathcal{H},$$

then $\alpha(f_1 f_2) - \alpha(f_1)\alpha(f_2)$ is compact on $P\mathcal{H}$, for each $f_1, f_2 \in C(S^1)$. For $f(z) = z$, $\alpha(z)$ is a Fredholm operator, which means that it is invertible modulo compact operators and thus has a well defined Fredholm index given by

$$\mathrm{Index}(\alpha(z)) = \mathrm{Dim}(\ker \alpha(z)) - \mathrm{Dim}(\mathrm{coker}\, \alpha(z))$$

Show that the assignment $(U, \pi, F) \mapsto \mathrm{Index}(\alpha(z))$ extends to a well defined isomorphism $K_1(S^1) \to \mathbf{Z}$.

Exercise 4.2.19. Show that $K_1([0,1]) = 0$ (without using the homotopy invariance).

Chapter 5

Kasparov's Equivariant KK-theory

In a series of papers [45] [46] [47] [50] from 1980 to 1988, G. Kasparov defined an *equivariant KK-theory* for pairs of C*-algebras, a powerful machinery to deal both with K-theory and K-homology of C*-algebras.

For a discrete group Γ, a Γ-*C*-algebra* is a C*-algebra endowed with an action of Γ by *-automorphisms. To any pair (A, B) of Γ-C*-algebras, G. Kasparov associates two abelian groups

$$KK_i^\Gamma(A, B) \qquad (i = 0, 1)$$

that we shall discuss in this chapter. For a C*-algebra A, a *Hilbert C*-module* over A is a right A-module $\mathcal{E}$ equipped with an A-valued scalar product, namely a map

$$\langle\ ,\ \rangle_A : \mathcal{E} \times \mathcal{E} \to A$$

that satisfies for all $x, x' \in \mathcal{E}$ and $a \in A$:

$$\begin{aligned}
\langle x, y\rangle_A &= \langle y, x\rangle_A^* \\
\langle x + x', y\rangle_A &= \langle x, y\rangle_A + \langle x', y\rangle_A \\
\langle x, ya\rangle_A &= \langle x, y\rangle_A\, a \\
\langle x, x\rangle_A &\geq 0 \text{ with equality if and only if } x = 0;
\end{aligned}$$

moreover $\mathcal{E}$ has to be complete with respect to the norm

$$\|x\|_{\mathcal{E}} = \|\, \langle x, x\rangle_A \,\|^{1/2}.$$

For a good reference for Hilbert C*-modules see [60].

Example 5.1. We endow $\mathcal{H}_A = \ell^2(\mathbf{N})\otimes A$ with an A-valued scalar product defined as follows:

$$\langle\xi|\eta\rangle_A = \sum_{n\in\mathbf{N}} \xi_n^*\eta_n$$

for all $\xi,\eta\in\mathcal{H}_A$. This turns $\mathcal{H}_A$ into a Hilbert C*-module over A.

Definition 5.2. Let $\mathcal{E}$ be a Hilbert C*-module over A. Set

$$\mathcal{L}_A(\mathcal{E})=\{T\colon\mathcal{E}\to\mathcal{E}\mid\exists\, T^*\colon\mathcal{E}\to\mathcal{E} \text{ such that, for every } \xi,\eta\in\mathcal{E}\colon \langle T\xi|\eta\rangle_A=\langle\xi|T^*\eta\rangle_A\}$$

Operators in $\mathcal{L}_A(\mathcal{E})$ are automatically A-linear and continuous. For the operator norm, $\mathcal{L}_A(\mathcal{E})$ is a C*-algebra, which plays the role of continuous linear operators on a Hilbert space. An operator $T\in\mathcal{L}_A(\mathcal{E})$ is *compact in the sense of C*-modules* if it is a norm limit of finite rank operators; and it is *finite rank in the sense of C*-modules* if it is a linear combination of operators of the form $\theta_{\xi,\eta}$, where $\xi,\eta\in\mathcal{E}$ and

$$\theta_{\xi,\eta}(x) = \xi\,\langle\eta,x\rangle_A \qquad (x\in\mathcal{E}).$$

Exercise 5.3. Show that an operator T on $\mathcal{H}_A$ is compact in the sense of C*-modules if, and only if

$$\lim_{n\to\infty} \|T|_{(A^n)^\perp}\| = 0.$$

Definition 5.4 (G. Kasparov). A *cycle over* (A,B) is a triple $(U,\pi,\mathcal{F})$ where:

- U is a representation of Γ on some Hilbert C*-module $\mathcal{E}$ over B, *unitary* in the sense that
$$\langle U_\gamma\xi|U_\gamma\eta\rangle_B = \gamma\cdot\langle\xi|\eta\rangle_B\,.$$
for all $\gamma\in\Gamma$ and $\xi,\eta\in\mathcal{E}$.
- $\pi : A\to\mathcal{L}_B(\mathcal{E})$ is a *-homomorphism which is *covariant*, in the sense that
$$U_\gamma\pi(a)U_{\gamma^{-1}} = \pi(\gamma\cdot a).$$
for all $\gamma\in\Gamma$ and $a\in A$.
- $\mathcal{F}$ is a self-adjoint operator in $\mathcal{L}_B(\mathcal{E})$.

Moreover we require the following operators

$$\pi(a)(\mathcal{F}^2-1)\,,\ [\pi(a),\mathcal{F}]\,,\ [U_\gamma,\mathcal{F}] \ \text{ for all } a\in A\,,\gamma\in\Gamma$$

to be compact in the sense of C*-modules.

Definition 5.5. The distinction between *even* and *odd* cycles is made by requiring the situation to be $\mathbf{Z}_2$-graded in the even case and ungraded in the odd case.

A cycle is *degenerate* if the operators $\pi(a)(\mathcal{F}^2-1)$, $[\pi(a),\mathcal{F}]$ and $[U_\gamma,\mathcal{F}]$ are zero for all $a\in A$ and $\gamma\in\Gamma$.

Two cycles $\alpha_0 = (U_0, \pi_0, \mathcal{F}_0)$ and $\alpha_1 = (U_1, \pi_1, \mathcal{F}_1)$ are *homotopic* if $U_0 = U_1$, $\pi_0 = \pi_1$ and there exists a norm continuous path $(\mathcal{F}_t)_{t\in[0,1]}$, connecting $\mathcal{F}_0$ to $\mathcal{F}_1$ and such that for each $t \in [0,1]$ the triple $\alpha_t = (U_0, \pi_0, \mathcal{F}_t)$ is a cycle (of the same parity).

Two cycles α_0 and α_1 are *equivalent* and we write $\alpha_0 \sim \alpha_1$ if there exists two degenerate cycles β_0 and β_1 such that, up to unitary equivalence, $\alpha_0 \oplus \beta_0$ is homotopic to $\alpha_1 \oplus \beta_1$.

We will write $KK_0^\Gamma(A,B)$ for the set of equivalence classes of even cycles over (A,B) and $K_1^\Gamma(A,B)$ for the set of equivalence classes of odd cycles over (A,B). These are in fact abelian groups (the proof is the same as in Proposition 4.2.7).

In case where Γ is the trivial group, we just write $KK_0(A,B)$ and $KK_1(A,B)$.

Kasparov's stabilization Theorem [46] allows one to assume $\mathcal{E} = \mathcal{H}_B$ in the definition of cycles over (A,B): it says that, if $\mathcal{E}$ is a separable Hilbert C*-module over the C*-algebra B, then there exists a Hilbert C*-module $\mathcal{E}'$ over B such that

$$\mathcal{E} \oplus \mathcal{E}' \simeq \mathcal{H}_B.$$

Notice that in doing so the representation of A doesn't necessarily remain essential.

Remark 5.6. The functors KK_i^Γ for $i = 0,1$ are covariant in B and contravariant in A. Indeed, let $\alpha = (U, \pi, \mathcal{F}) \in KK_i^\Gamma(A,B)$ be a cycle and consider a *-homomorphism $\theta : C \to A$, it defines

$$\theta^*\alpha = (U, \pi \circ \theta, \mathcal{F}) \in KK_i^\Gamma(C,B),$$

and if $\theta : B \to C$ is a *-homomorphism, consider the Hilbert C*-module over C given by $\mathcal{E} \otimes_B C$ and define

$$\theta_*\alpha = (U \otimes 1, \pi \otimes 1, \mathcal{F} \otimes 1) \in KK_i^\Gamma(A,C).$$

Example 5.7. (1) Let Γ be a discrete group and X a locally compact proper Γ-space. Then

$$K_i^\Gamma(X) = KK_i^\Gamma(C_0(X), \mathbf{C}) \qquad (i = 0,1)$$

since in the case of proper actions we may always assume that $[U_\gamma, \mathcal{F}] = 0$ for all $\gamma \in \Gamma$ (see the proof of Lemma 6.1.2 below for that).

(2) If B is any C*-algebra, then (taking Γ as to be the trivial group)

$$KK_i(\mathbf{C}, B) \simeq K_i(B) \qquad (i = 0,1)$$

where $K_i(B)$ is the K-theory of B as defined in Chapter 3. We will now give the maps that realize the above isomorphisms, in case B is unital.

<u>$i = 0$</u> : If $x = [e_0] - [e_1] \in K_0(B)$ with e_0 and e_1 idempotent matrices in $M_n(B)$, consider $e_0 \oplus 0$ and $e_1 \oplus 0$ as operators acting on $\mathcal{H}_B$. For $\lambda \in \mathbf{C}$, define

$$\pi(\lambda) = \begin{pmatrix} \lambda(e_0 \oplus 0) & 0 \\ 0 & \lambda(e_1 \oplus 0) \end{pmatrix}$$

in the $\mathbf{Z}_2$-graded C*-module $\mathcal{H}_B \oplus \mathcal{H}_B$ and take $\mathcal{F} = 0$ (thus $[\pi(\lambda), \mathcal{F}] = 0$ for all $\lambda \in \mathbf{C}$). Since π is a representation by compact operators

$$\pi(\lambda)(\mathcal{F}^2 - 1) = -\pi(\lambda)$$

is also compact. We then define the map $K_0(B) \to KK_0(\mathbf{C}, B)$ by associating to an element $x = [e_0] - [e_1] \in K_0(B)$ the cycle $(U, \pi, \mathcal{F})$ as defined above. For the proof that this map is an isomorphism we refer to V. Lafforgue's thesis [59].

$\underline{i=1}$: Consider the group $\mathcal{G}$ of invertible operators $\mathcal{V}$ on $\mathcal{H}_B$ such that $\mathcal{V} - 1$ is compact. Then $K_1(B) \simeq \pi_0(\mathcal{G})$, the map $K_1(B) \to \pi_0(\mathcal{G})$ being induced by

$$S \mapsto S \oplus 1 = \begin{pmatrix} S & 0 \\ 0 & 1 \end{pmatrix},$$

for $S \in GL_n(B)$. To define a map $KK_1(\mathbf{C}, B) \to K_1(B)$, start with $\mathcal{F} = \mathcal{F}^*$ on a Hilbert C*-module $\mathcal{E}$ such that $\mathcal{F}^2 - 1$ is compact. By the stabilization Theorem, we may assume that $\mathcal{E} = \mathcal{H}_B$. Then

$$-\exp(i\pi\mathcal{F}) \in \mathcal{G},$$

and in this way we get a map $KK_1(\mathbf{C}, B) \to \pi_0(\mathcal{G}) \simeq K_1(B)$.

(3) Take $\mathcal{E} = C_0(\mathbf{R})$ as a Hilbert C*-module over itself, and $\mathcal{F}$ the multiplication by $x \mapsto \dfrac{x}{\sqrt{1+x^2}}$. Then $(\mathcal{E}, \mathcal{F})$ is an element of $KK_1(\mathbf{C}, C_0(\mathbf{R}))$ which, under the isomorphism

$$KK_1(\mathbf{C}, C_0(\mathbf{R})) \to K_1(C_0(\mathbf{R})),$$

goes to a generator of $K_1(C_0(\mathbf{R})) \simeq \mathbf{Z}$.

(4) Identify $\mathbf{R}^2$ with $\mathbf{C}$ and consider $\mathcal{E} = C_0(\mathbf{C}) \oplus C_0(\mathbf{C})$ as a $\mathbf{Z}_2$-graded Hilbert C*-module over $C_0(\mathbf{R}^2)$, and $\mathcal{F} = \begin{pmatrix} 0 & P^* \\ P & 0 \end{pmatrix}$ where P is the multiplication by $z \mapsto \dfrac{z}{\sqrt{1+|z|^2}}$. Then $(\mathcal{E}, \mathcal{F})$ is an element of $KK_1(\mathbf{C}, C_0(\mathbf{R}^2))$ which, under the isomorphism

$$KK_1(\mathbf{C}, C_0(\mathbf{R}^2)) \to K_1(C_0(\mathbf{R}^2)),$$

goes to a generator of $K_1(C_0(\mathbf{R}^2)) \simeq \mathbf{Z}$.

(5) Suppose that $\theta : A \to B$ is a Γ-equivariant *-homomorphism, it defines a class $[\theta] \in KK_0^\Gamma(A, B)$. Indeed, view B as a Hilbert C*-module over itself, the representation U of the group Γ will be given by

$$U_\gamma \xi = \gamma(\xi)$$

on B, so that we can consider $U \oplus 0$ on the $\mathbf{Z}_2$-graded Hilbert C*-module $B \oplus 0$. The representation π of A will be given by $\pi = \pi^0 \oplus 0$ where for $a \in A$, $\pi^0(a) = \theta(a)$. We take $F = 0$ on $B \oplus 0$.

Theorem 5.8 (G. Kasparov, [45] and [50]). *Let us restrict to separable C^*-algebras. Let A, B, C be Γ-C^*-algebras. Then there is a bi-additive pairing, for $i, j, i+j \in \mathbf{Z}_2$*

$$\begin{array}{rcl} KK_i^\Gamma(A,B) \times KK_j^\Gamma(B,C) & \to & KK_{i+j}^\Gamma(A,C) \\ (\ x\ ,\ y\ ) & \mapsto & x \otimes_B y \end{array}$$

the Kasparov product, *which is associative and functorial in all possible senses. Moreover, for any separable Γ-C^*-algebra D there is a homomorphism of* extension of scalars

$$\begin{array}{rcl} \tau_D : KK_i^\Gamma(A,B) & \to & KK_i^\Gamma(A \otimes D, B \otimes D) \\ (U, \pi, \mathcal{F}) & \mapsto & (U \otimes 1, \pi \otimes 1, \mathcal{F} \otimes 1) \end{array}$$

and a descent homomorphism

$$j_\Gamma : KK_i^\Gamma(A,B) \to KK_i(A \rtimes_r \Gamma, B \rtimes_r \Gamma).$$

Both are functorial in all possible senses.

Remarks 5.9. (a) In particular we have that:

- if $\alpha : A \to B$ is a Γ-equivariant $*$-homomorphism and y an element in $KK_j^\Gamma(B,C)$, then

$$[\alpha] \otimes_B y = \alpha^*(y) \in KK_j^\Gamma(A,C),$$

- if $\beta : B \to C$ is a Γ-equivariant $*$-homomorphism and x an element in $KK_i^\Gamma(A,B)$, then

$$x \otimes_B [\beta] = \beta_*(x) \in KK_i^\Gamma(A,C).$$

(b) If $x = (U, \pi, \mathcal{F})$ and $y = (V, \rho, \mathcal{G})$ are cycles over (A,B) and (B,C) respectively. Denote by $\mathcal{E}_B$ (resp. $\mathcal{E}_C$) the underlying Hilbert B-module for x (resp. C-module for y), then $\mathcal{E} = \mathcal{E}_B \widehat{\otimes}_B \mathcal{E}_C$ (B acts on $\mathcal{E}_C$ via ρ) is a Hilbert C-module, and

$$x \otimes_B y = (U \otimes V, \pi \otimes 1, \mathcal{H}).$$

For the definition of the operator $\mathcal{H}$ see [50] and notice that taking $\mathcal{F} \otimes \mathcal{G}$ wouldn't make sense since $\mathcal{G}$ doesn't exactly commute with ρ.
(c) Given $x \in KK_i^\Gamma(A,B)$ and $y \in KK_0^\Gamma(\mathbf{C},\mathbf{C})$, we have that

$$\tau_A(y) \otimes_A x = x \otimes_B \tau_B(y)$$

where the homomorphisms

$$\tau_A : KK_0^\Gamma(\mathbf{C},\mathbf{C}) \to KK_0^\Gamma(A,A) \quad \text{and} \quad \tau_B : KK_0^\Gamma(\mathbf{C},\mathbf{C}) \to KK_0^\Gamma(B,B)$$

are the extension of scalars associated to A and B respectively. In particular (taking $A = B = \mathbf{C}$), we see that $KK_0^\Gamma(\mathbf{C}, \mathbf{C})$ is a ring, which is commutative and unital (with unit $1 = [\mathrm{id}_{\mathbf{C}}]$). All groups $KK_i^\Gamma(A, B)$ and $KK_i(A \rtimes_r \Gamma, B \rtimes_r \Gamma)$ are modules over that ring, which plays a big role for that reason.

(d) We will give the descent homomorphism j_Γ explicitly. Take

$$(U, \pi, \mathcal{F}) \in KK_i^\Gamma(A, B),$$

with underlying B-module $\mathcal{E}_B$, and define $\widetilde{\mathcal{E}} = \mathcal{E}_B \otimes \mathbf{C}\Gamma = C(\Gamma, \mathcal{E}_B)$, this is a $B \rtimes_r \Gamma$-module with $B \rtimes_r \Gamma$-scalar product given by

$$\langle \xi, \eta \rangle (\gamma) = \sum_{s \in \Gamma} s \left\langle \xi(s), \eta(s^{-1}\gamma) \right\rangle_B$$

Define, for $a \in C(\Gamma, A)$, $\xi \in \widetilde{\mathcal{E}}$ and $\gamma \in \Gamma$:

$$(\widetilde{\pi}(a)\xi)(\gamma) = \sum_{s \in \Gamma} \pi\left(a(s)\right) U_s \left(\xi(s^{-1}\gamma)\right)$$

and $(\widetilde{\mathcal{F}}\xi)(\gamma) = \mathcal{F}\left(\xi(\gamma)\right)$, so that we finally set $j_\Gamma(U, \pi, \mathcal{F}) = (\widetilde{\pi}, \widetilde{\mathcal{F}})$.

Chapter 6

The Analytical Assembly Map

To illustrate the difficulty of constructing an assembly map, consider the following situation. Let X be a proper Γ-compact space and

$$(U, \pi, F = \begin{pmatrix} 0 & P^* \\ P & 0 \end{pmatrix})$$

an even cycle in $K_0^\Gamma(X)$. The goal is to define, out of these data, an element in $K_0(C_r^*\Gamma)$. A naive approach would be to consider the kernel and co-kernel of P: these are indeed modules over $C_r^*\Gamma$, but these modules are in general not projective of finite type, as shown in the following example:

Let $\Gamma = \mathbf{Z}$, $X = \mathbf{Z}$. Consider the cycle (U, π, F) where U is the left regular representation on $\mathcal{H} = \ell^2\mathbf{Z}$, π the representation of $C_0(\mathbf{Z})$ by pointwise multiplication and $P = 0$. By Example 4.2.3 (1), (U, π, F) is an even cycle, but $\ker(P) = \ell^2\mathbf{Z}$ is not projective of finite type as a $C_r^*\Gamma$-module. Indeed, via Fourier transform, it gives $L^2(S^1)$ as a module over $C(S^1)$ acting by pointwise multiplication.

We shall give 2 approaches to construct μ_i^Γ.

6.1 First approach: à la Baum-Connes-Higson

We follow here the construction in [9].

Definition 6.1.1. Let π be a representation of $C_0(X)$ on a Hilbert space $\mathcal{H}$. An operator $T \in \mathcal{B}(\mathcal{H})$ is *properly supported* (with respect to π) if, for every $f \in C_c(X)$, there exists $g \in C_c(X)$ such that

$$T\pi(f) = \pi(g)T\pi(f)$$

(this expresses a locality condition for the operator T).

Lemma 6.1.2. *Let X be a proper Γ-compact space and $(U,\pi,F) \in K_i^\Gamma(X)$ be a cycle. Then (U,π,F) is homotopic to a cycle (U,π,F') with F' properly supported.*

Proof. Without loss of generality we may assume that the representation π is essential (see Exercise 4.2.16). Fix $h \in C_c(X)$, $h \geq 0$ and that satisfies, for every $x \in X$: $\sum_{\gamma\in\Gamma} h(\gamma^{-1}x) = 1$. Such an h can be built starting with a $g \in C_c(X)$, $g \geq 0$ and such that $\Gamma x \cap \operatorname{supp}(g) \neq \emptyset$ for every $x \in X$ (such a g exists because of Γ-compactness) and then setting

$$h(x) = \frac{g(x)}{\sum_{\gamma\in\Gamma} g(\gamma x)}.$$

Since $F = F^*$, we have, for every $\gamma \in \Gamma$ the operator inequalities:

$$\begin{aligned} -U_\gamma \pi(h) U_{\gamma^{-1}} \|F\| &\leq U_\gamma \pi(\sqrt{h}) F \pi(\sqrt{h}) U_{\gamma^{-1}} \\ &\leq U_\gamma \pi(h) U_{\gamma^{-1}} \|F\|. \end{aligned}$$

Summing over Γ and setting

$$F' = \sum_{\gamma\in\Gamma} U_\gamma \pi(\sqrt{h}) F \pi(\sqrt{h}) U_{\gamma^{-1}},$$

and since by covariance $U_\gamma \pi(h) U_{\gamma^{-1}} = \pi(\gamma \cdot h)$ (where $\gamma \cdot h(x) = h(\gamma^{-1}x)$), we get

$$-\sum_{\gamma\in\Gamma} \pi(\gamma\cdot h)\|F\| \leq F' \leq \sum_{\gamma\in\Gamma} \pi(\gamma\cdot h)\|F\|.$$

Since $\sum_{\gamma\in\Gamma} \gamma \cdot h = 1$, we see that the sum defining F' converges in the strong topology, and that $\|F'\| \leq \|F\|$. Note that F' is trivially Γ-invariant. For $f \in C_c(X)$, we have

$$F'\pi(f) = \sum_{\gamma\in\Gamma} U_\gamma \pi(\sqrt{h}) F \pi\left(\sqrt{h}(\gamma^{-1}\cdot f)\right) U_{\gamma^{-1}}.$$

Since the action of Γ on X is proper, the set $S = \{\gamma \in \Gamma : \sqrt{h}(\gamma^{-1}\cdot f) \neq 0\}$ is finite, so that the preceding sum is a finite one. Now, pick a $g \in C_c(X)$ which is 1 on $\bigcup_{\gamma\in S} \gamma(\operatorname{supp}(h))$; then $F'\pi(f) = \pi(g)F'\pi(f)$, i.e. F' is properly supported. Again since $\sum_{\gamma\in\Gamma} \gamma\cdot h = 1$, we have

$$\begin{aligned} F - F' &= \sum_{\gamma\in\Gamma} \left(U_\gamma \pi(h) U_{\gamma^{-1}} F - U_\gamma \pi(\sqrt{h}) F \pi(\sqrt{h}) U_{\gamma^{-1}} \right) \\ &= \sum_{\gamma\in\Gamma} U_\gamma \pi(\sqrt{h}) [\pi(\sqrt{h}) U_{\gamma^{-1}}, F] \\ &= \sum_{\gamma\in\Gamma} U_\gamma \pi(\sqrt{h}) [\pi(\sqrt{h}), F] U_{\gamma^{-1}} \end{aligned}$$

and by assumption each term in this summation is compact. Now, for $f \in C_c(X)$:

$$\pi(f)(F-F') = \sum_{\gamma\in\Gamma} U_\gamma \pi\left((\gamma^{-1}\cdot f)\sqrt{h}\right)[\pi(\sqrt{h}), F]U_{\gamma^{-1}}$$

and this summation is indexed by the same finite set S as above, so that $\pi(f)(F-F')$ is a compact operator and therefore (U,π,F') defines a cycle homotopic to (U,π,F). □

Now, given a cycle $(U,\pi,F) \in K_i^\Gamma(X)$, we may assume that F is properly supported and π is essential. We then form the dense subspace

$$H = \pi(C_c(X))\mathcal{H}$$

and notice that $F(H) \subseteq H$. We view H as a right $\mathbf{C}\Gamma$-module by

$$\xi\cdot\gamma = U_{\gamma^{-1}}\xi$$

for all $\xi \in H$ and $\gamma \in \Gamma$, and define a $\mathbf{C}\Gamma$-valued scalar product on H by

$$\langle \xi_1, \xi_2\rangle(\gamma) = \langle \xi_1, U_\gamma \xi_2\rangle_{\mathcal{H}}$$

(note that $\gamma \mapsto \langle \xi_1, \xi_2\rangle(\gamma)$ has finite support because Γ acts properly on X). For $\xi \in H$, let us show that the function

$$\varphi : \gamma \mapsto \langle \xi,\xi\rangle(\gamma) = \langle \xi, U_\gamma\xi\rangle$$

defines a positive element in $C_r^*\Gamma$; this function is positive definite on Γ, meaning that, for every $f \in \mathbf{C}\Gamma$:

$$\sum_{s,t\in\Gamma} \overline{f(s)}f(t)\varphi(s^{-1}t) \geq 0,$$

which can be rewritten $\langle \check{f}, \lambda_\Gamma(\varphi)\check{f}\rangle \geq 0$ for every $f \in \mathbf{C}\Gamma$ (where $\check{f}(\gamma) = f(\gamma^{-1})$). But since $\mathbf{C}\Gamma$ is dense in $\ell^2\Gamma$, this also holds for every $f \in \ell^2\Gamma$, showing that $\lambda_\Gamma(\varphi)$ is a positive operator. So we complete H into a Hilbert C*-module $\mathcal{E}$ over $C_r^*\Gamma$, with respect to this $\mathbf{C}\Gamma$-valued scalar product.

Lemma 6.1.3. *The operator F extends continuously to a bounded operator $\mathcal{F}$ on $\mathcal{E}$.*

Proof. We notice first that, for $\xi \in H$, the function

$$\gamma \mapsto \|F\|^2\langle \xi, U_\gamma\xi\rangle - \langle F(\xi), U_\gamma F(\xi)\rangle$$

is positive-definite on H. Indeed, for $f \in \mathbf{C}\Gamma$:

$$\begin{aligned}
&\sum_{s,t\in\Gamma} \overline{f(s)}f(t)\left(\|F\|^2\langle \xi, U_{st^{-1}}\xi\rangle - \langle F(\xi), U_{st^{-1}}F(\xi)\rangle\right)\\
&= \|F\|^2\|\sum_{s\in\Gamma} f(s)U_{s^{-1}}\xi\|^2 - \|F(\sum_{s\in\Gamma} f(s)U_{s^{-1}}\xi)\|^2 \geq 0
\end{aligned}$$

(the equality uses $U_\gamma F = FU_\gamma$). Therefore

$$\langle F\xi, F\xi\rangle(\cdot) \leq \|F\|^2 \langle \xi,\xi\rangle(\cdot)$$

in the sense of order in the C*-algebra $C_r^*\Gamma$, hence

$$\|F\xi\|_{\mathcal{E}} \leq \|F\| \, \|\xi\|_{\mathcal{E}}$$

and thus F extends continuously to a self-adjoint operator $\mathcal{F}$ on $\mathcal{E}$, such that $\|\mathcal{F}\| \leq \|F\|$. This proves the lemma. □

The crucial property of $\mathcal{F}$ is:

Proposition 6.1.4. *The operator $\mathcal{F}^2 - 1$ on the Hilbert C^*-module $\mathcal{E}$ is compact.*

For a proof, we refer to [88]. This proposition says that $\mathcal{F}$ defines an element in $KK_i(\mathbf{C}, C_r^*\Gamma) = K_i(C_r^*\Gamma)$. So we set

$$\mu_i^\Gamma(U,\pi,F) = [\mathcal{F}];$$

this map extends continuously to the direct limit and defines the assembly map:

$$\mu_i^\Gamma : RK_i^\Gamma(\underline{E}\Gamma) \to K_i(C_r^*\Gamma). \qquad (i = 0,1)$$

Example 6.1.5. Let H be a finite subgroup of Γ, $X = \Gamma/H$ and σ a finite dimensional representation of H, with space V_σ. By Example 4.2.3 (1), this induces a cycle $\beta_{H,\sigma}$ in $K_0^\Gamma(X)$ with underlying Hilbert space

$$\mathcal{H} = \mathrm{Ind}_H^\Gamma \sigma,$$

the space of the induced representation. Suppose that σ is irreducible, so that $V_\sigma \simeq \mathbf{C}H\overline{p_\sigma}$ (where p_σ is some projection in $\mathbf{C}H$). Then

$$\mathcal{H} = \ell^2\Gamma \otimes_{\mathbf{C}H} V_\sigma$$

and $\pi(C_c(X))\mathcal{H} = \mathbf{C}\Gamma \otimes_{\mathbf{C}H} V_\sigma = \mathbf{C}\Gamma\overline{p_\sigma}$. Then, for $\xi_1,\xi_2 \in \mathbf{C}\Gamma\overline{p_\sigma}$:

$$\langle \xi_1,\xi_2\rangle(\gamma) = \langle \xi_1, \lambda_\Gamma(\gamma)\xi_2\rangle = (\overline{\xi}_1 * \check{\xi}_2)(\gamma) = (\check{\xi}_1^* * \check{\xi}_2)(\gamma)$$

Using $\check{\overline{p_\sigma}} = {p_\sigma}^* = p_\sigma$, this shows that the map

$$\begin{array}{ccc} \mathbf{C}\Gamma\overline{p_\sigma} & \to & p_\sigma\mathbf{C}\Gamma \\ \xi & \mapsto & \check{\xi} \end{array}$$

extends to an isometric C*-module isomorphism from the completion $\mathcal{E}$ of $\mathbf{C}\Gamma\overline{p_\sigma}$, to the right ideal $p_\sigma C_r^*\Gamma$ in $C_r^*\Gamma$. In more down-to-earth language, we have

$$\mu_0^\Gamma(\beta_{H,\sigma}) = [p_\sigma],$$

the K-theory class of p_σ viewed as a projection in $C_r^*\Gamma$.

In particular, for $H = 1$ and σ the trivial 1 dimensional representation, then $\beta_{H,\sigma}$ is nothing but the element $[i_*] \in RK_0(\underline{E}\Gamma)$ corresponding to the inclusion of the base-point in $B\Gamma$, so that

$$\mu_0^\Gamma[i_*] = [1].$$

Suppose that Γ is a finite group. In this case both $K_0^\Gamma(\mathrm{pt})$ and $K_0(C_r^*\Gamma)$ are abstractly isomorphic to the additive group of the representation ring $R(\Gamma)$, i.e. to the free abelian group on the set $\hat{\Gamma}$ of isomorphism classes of irreducible representations of Γ. In the above construction, take $H = \Gamma$ and let σ run along $\hat{\Gamma}$. Then the $\beta_{H,\sigma}$'s run along a set of generators of $K_0^\Gamma(\mathrm{pt})$ and the $[p_{\Gamma,\sigma}]$'s run along a set of generators of $K_0(C_r^*\Gamma)$. On the other hand:

$$K_1^\Gamma(\underline{E}\Gamma) = 0 = K_1(C_r^*\Gamma)$$

for Γ finite. In other words, we have checked that the Baum-Connes Conjecture holds for finite groups.

Example 6.1.6. Let us come back to Example 4.2.3 (3), i.e. the generator of $K_1^{\mathbf{Z}}(\mathbf{R}) \simeq \mathbf{Z}$. It is given by the triple (U, π, F) where U is the representation of $\mathbf{Z}$ on $L^2(\mathbf{R})$ by integer translations, π is the representation of $C_0(\mathbf{R})$ by pointwise multiplication and F is the Hilbert transform (note that F is not properly supported). Fourier transforming this triple, we get the triple (V, ρ, G), where V_n is the pointwise multiplication by

$$\lambda \mapsto e^{-2\pi i n\lambda}$$

on $L^2(\hat{\mathbf{R}})$ (and $n \in \mathbf{Z}$), ρ is the representation of $C_0(\mathbf{R})$ by convolution by Fourier transforms, and G is the pointwise multiplication by

$$\lambda \mapsto \mathrm{sgn}(\lambda).$$

By means of the homotopy

$$(t, \lambda) \mapsto \cos(\frac{\pi t}{2})\mathrm{sgn}(\lambda) + \sin(\frac{\pi t}{2})\frac{\lambda}{\sqrt{1+\lambda^2}} \qquad (t \in [0,1],\ \lambda \in \hat{\mathbf{R}})$$

we may assume that G is pointwise multiplication by

$$f(\lambda) = \frac{\lambda}{\sqrt{1+\lambda^2}}.$$

Note then that the Schwartz space $\mathcal{S}(\hat{\mathbf{R}})$ is a common invariant subspace for V, ρ and G. We want to perform on $\mathcal{S}(\hat{\mathbf{R}})$ the construction following Lemma 6.1.2, i.e. to complete it into a Hilbert C*-module over $C_r^*\mathbf{Z} \simeq C(S^1)$. First we turn $\mathcal{S}(\hat{\mathbf{R}})$ into a right $\mathbf{CZ}$-module by setting

$$(\xi \cdot n)(\lambda) = e^{2\pi i n\lambda}\xi(\lambda)$$

for $\xi \in \mathcal{S}(\hat{\mathbf{R}}), n \in \mathbf{Z}, \lambda \in \hat{R}$. The $C_r^*\mathbf{Z}$-valued scalar product on $\mathcal{S}(\hat{\mathbf{R}})$ is given by

$$\langle \xi_1, \xi_2 \rangle_n = \langle \xi_1, V_n \xi_2 \rangle = \int_{-\infty}^{+\infty} \overline{\xi_1(\lambda)} e^{-2\pi i n \lambda} \xi_2(\lambda) d\lambda,$$

for $\xi_1, \xi_2 \in \mathcal{S}(\hat{\mathbf{R}}), n \in \mathbf{Z}$. Under the identification $C_r^*\mathbf{Z} \simeq C(S^1)$ given by Fourier series, namely

$$a \mapsto \{\theta \mapsto \sum_{n \in \mathbf{Z}} a(n) e^{2\pi i n \theta}\}$$

for $a \in \mathbf{C}\mathbf{Z}$, $\theta \in S^1$, this becomes a $C(S^1)$-valued scalar product

$$\langle \xi_1, \xi_2 \rangle (\theta) = \sum_{n \in \mathbf{Z}} \int_{-\infty}^{+\infty} \overline{\xi_1(\lambda)} \xi_2(\lambda) e^{2\pi i n(\theta - \lambda)} d\lambda.$$

Let $\mathcal{E}$ be the completion of $\mathcal{S}(\hat{\mathbf{R}})$ for this $C(S^1)$-valued scalar product; let $\mathcal{F}$ be the continuous extension of G to $\mathcal{E}$ given by Lemma 6.1.3. According to Proposition 6.1.4, $\mu_1^{\mathbf{Z}}(V, \rho, G)$ is given by $[\mathcal{F}] \in KK_1(\mathbf{C}, C_r^*\mathbf{Z}) \simeq K^1(S^1)$. We wish to give a more geometric description of the underlying Hilbert C*-module $\mathcal{E}$.

To do that, let $\mathbf{Z}$ act on $\hat{\mathbf{R}}$ by integer translations, and consider the Hilbert bundle over $S^1 = \hat{\mathbf{R}}/\mathbf{Z}$ induced by the left regular representation of $\mathbf{Z}$; the total space of this bundle is $\hat{\mathbf{R}} \times_{\mathbf{Z}} \ell^2\mathbf{Z}$ and its space of sections is

$$\mathcal{E}' = \{\eta : \hat{\mathbf{R}} \to \ell^2\mathbf{Z} |\ \eta \text{ continuous},\ \eta(\lambda+1)_n = \eta(\lambda)_{n+1}\ \forall\ \lambda \in \hat{\mathbf{R}},\ n \in \mathbf{Z}\}.$$

Note that a section is determined by its values on $[-1/2, 1/2[$. The action of $C(S^1)$ on $\mathcal{E}'$ is by pointwise multiplication. Moreover, pointwise scalar product turns $\mathcal{E}'$ into a $C(S^1)$-module:

$$\langle \eta_1, \eta_2 \rangle (\theta) = \sum_{n \in \mathbf{Z}} \overline{\eta_1(\tilde{\theta})_n} \eta_2(\tilde{\theta})_n$$

($\eta_1, \eta_2 \in \mathcal{E}'$; $\tilde{\theta}$ a real number that lifts $\theta \in S^1$). Consider the map $\psi : \mathcal{S}(\hat{\mathbf{R}}) \to \mathcal{E}'$ defined by

$$((\psi(\xi))(\lambda))_n = \xi(\lambda + n)$$

for $\xi \in \mathcal{S}(\hat{\mathbf{R}})$, $\lambda \in \hat{\mathbf{R}}$, $n \in \mathbf{Z}$. It is clear that ψ is a $\mathbf{C}\mathbf{Z}$-module map. Furthermore ψ is isometric with respect to the $C(S^1)$-valued scalar product, since:

$$\begin{aligned}
\langle \psi(\xi_1), \psi(\xi_2) \rangle (\theta) &= \sum_{n \in \mathbf{Z}} \overline{\xi_1(\tilde{\theta} + n)} \xi_2(\tilde{\theta} + n) = \sum_{n \in \mathbf{Z}} (\overline{\xi_1} \xi_2)(\tilde{\theta} + n) \\
&= \sum_{n \in \mathbf{Z}} \widehat{\overline{\xi_1}\xi_2}(n) e^{2\pi i n \theta} (\text{Poisson summation formula}) \\
&= \sum_{n \in \mathbf{Z}} \int_{-\infty}^{+\infty} \overline{\xi_1(\lambda)} e^{2\pi i n(-\lambda + \theta)} \xi_2(\lambda) d\lambda = \langle \xi_1, \xi_2 \rangle (\theta)
\end{aligned}$$

It is clear that ψ has dense image since, on the suitable subspace of smooth sections of $\hat{\mathbf{R}} \times_{\mathbf{Z}} \ell^2\mathbf{Z}$ with rapid decay in the fibers, ψ can be inverted by setting

$$\left(\psi^{-1}(\eta)\right)(\lambda) = \eta(\lambda)_0. \qquad (\eta \in \mathcal{E}',\ \lambda \in \hat{\mathbf{R}})$$

So ψ extends to an isometric isomorphism of Hilbert C*-modules over $C(S^1)$. Set $\mathcal{F}' = \psi\mathcal{F}\psi^{-1}$; then

$$(\mathcal{F}'\eta(\lambda))_n = f(\lambda+n)\eta(\lambda)_n.$$

Now we want to trace this element of $KK_1(\mathbf{C}, C(S^1))$ through the identification $KK_1(\mathbf{C}, C_r^*\mathbf{Z}) \simeq K_1(C(S^1))$ given in Example 5.7 (2). So, with $\mathcal{U} = -\exp(\pi i\mathcal{F}')$, we have

$$(\mathcal{U}\eta(\lambda))_n = -\eta(\lambda)_n \exp(i\pi f(\lambda+n)).$$

Set then

$$f_1(\lambda) = \begin{cases} -1 & \lambda \leq -1/2 \\ 2\lambda & -1/2 \leq \lambda \leq 1/2 \\ 1 & \lambda \geq 1/2 \end{cases}$$

and, for $s \in [0,1]$:

$$(\mathcal{U}_s\eta(\lambda))_n = -\eta(\lambda)_n \exp\Big(i\pi\left((1-s)f(\lambda+n) + sf_1(\lambda+n)\right)\Big).$$

Then $\mathcal{U}_s$ belongs to the group $\mathcal{G}$ of invertible operators on $\mathcal{E}'$ such that their difference with 1 is compact in the sense of Hilbert C*-modules over $C(S^1)$: here, this means that, for every $\lambda \in S^1$, the operator in the fiber $\ell^2\mathbf{Z}$ over λ is compact, and this operator depends norm-continuously on λ. Then, in $\pi_0(\mathcal{G}) \simeq K_1(C(S^1))$, we have $[\mathcal{U}] = [\mathcal{U}_1]$. But, for $\lambda \in [-1/2, 1/2[$

$$(\mathcal{U}_1\eta(\lambda))_n = \begin{cases} \eta(\lambda)_n & \text{for } n \neq 0 \\ -e^{2\pi i\lambda}\eta(\lambda)_0 & \text{for } n = 0. \end{cases}$$

The function $\lambda \mapsto -e^{2\pi i\lambda}$ gives the standard generator of $K_1(C(S^1)) = K^1(S^1) \simeq \mathbf{Z}$. Indeed, parametrizing S^1 with $[0,1[$ instead of $[-\frac{1}{2}, \frac{1}{2}[$, with $\mu = \lambda + \frac{1}{2}$ we get $\mu \mapsto e^{2\pi i\mu}$, as in Example 3.2.2 (3). This proves Conjecture 1 for the group $\mathbf{Z}$.

6.2 Second approach: à la Kasparov

Let X be a proper Γ-compact space. Fix $h \in C_c(X)$, $h \geq 0$ as in the first approach and let $e \in C_c(\Gamma \times X)$ be given by

$$e(\gamma, x) = \sqrt{h(x)h(\gamma^{-1}\cdot x)}$$

for each $(\gamma, x) \in \Gamma \times X$, then (using the product law in the crossed product $C_0(X) \rtimes_r \Gamma$, see Definition 2.3.1):

$$\begin{aligned} e^2(\gamma, x) &= \sum_{s\in\Gamma} e(s,x)e(s^{-1}\gamma, s^{-1}x) \\ &= \sum_{s\in\Gamma} \sqrt{h(x)h(s^{-1}x)h(s^{-1}x)h(\gamma^{-1}x)} \\ &= \sqrt{h(x)h(\gamma^{-1}x)} = e(\gamma, x) \end{aligned}$$

which means that $e = e^2 \in C_c(\Gamma \times X)$. This gives a class $[\mathcal{L}_X]$ in $K_0(C_0(X) \rtimes_r \Gamma)$, which clearly does not depend on the choice of the function h (the set of such functions is convex). On the other hand, we have Kasparov's descent homomorphism

$$j_\Gamma : K_i^\Gamma(X) = KK_i^\Gamma(C_0(X), \mathbf{C}) \to KK_i(C_0(X) \rtimes_r \Gamma, C_r^*\Gamma).$$

For $x \in K_i^\Gamma(X)$ we then define $\mu_i^\Gamma(x) \in K_i(C_r^*\Gamma)$ as the Kasparov product

$$\mu_i^\Gamma(x) = [\mathcal{L}_X] \otimes_{C_0(X)\rtimes_r\Gamma} j_\Gamma(x).$$

Note that it is possible to define μ_i^Γ without appealing to the Kasparov product; indeed, define the homomorphism $\theta : \mathbf{C} \to C_0(X) \rtimes_r \Gamma$ by

$$\theta(\lambda) = \lambda e. \qquad (\lambda \in \mathbf{C})$$

With the notations of Example 5.7 (5), we have that

$$[\theta] = [\mathcal{L}_X] \quad \text{in} \quad KK_0(\mathbf{C}, C_0(X) \rtimes_r \Gamma).$$

So, by Remark 5.9 (a), we could also define $\mu_i^\Gamma(x) = \theta^*(j_\Gamma(x))$. This bridges the two approaches.

Example 6.2.1. If the group Γ is finite and $X = \{\text{pt}\}$, then take

$$e(\gamma) = \frac{1}{|\Gamma|} \in \mathbf{C}\Gamma.$$

In any representation of Γ, this is the projection on the space of Γ-fixed points. If $F = \begin{pmatrix} 0 & P^* \\ P & 0 \end{pmatrix}$, where $P : \mathcal{H}_0 \to \mathcal{H}_1$ is a Fredholm operator, one finds $\mu_0^\Gamma(F) = [\ker(P)] - [\operatorname{coker}(P)] \in R(\Gamma)$, because of the following exercise.

Exercise 6.2.2. If π is any representation of Γ, with space V_π, then the right regular representation of Γ on $(V_\pi \otimes \mathbf{C}\Gamma)^\Gamma$ is equivalent to π. Together with the previous example, this statement re-proves Conjecture 1 for finite groups.

6.3 How to deduce the Kaplansky-Kadison conjecture

Here we show, as promised in Section 2.1, that for Γ a torsion free group, surjectivity of the Baum-Connes map μ_0^Γ implies the Kaplansky-Kadison conjecture on idempotents (that is, Conjecture 3). Recall that, for Γ torsion free, the left hand side of the Baum-Connes conjecture is $RK_0(B\Gamma) = \lim_{\to X} K_0(X)$, where X runs over compacts subsets of $B\Gamma$. Recall also that we denote by $\tau : C_r^*\Gamma \to \mathbf{C}$ the canonical trace, and by $\tau_* : K_0(C_r^*\Gamma) \to \mathbf{R}$ the induced map in K-theory (see Remark 3.1.10 and Exercise 3.1.11).

Proposition 6.3.1. *For Γ torsion free, the image of*

$$\tau_* \circ \mu_0^\Gamma : RK_0(B\Gamma) \to \mathbf{R}$$

is the group of integers.

Proof. Taking for $E\Gamma = \underline{E}\Gamma$ the model described in Example 4.1.8 (4), and $B\Gamma = E\Gamma/\Gamma$, we see that in the limit $RK_0(B\Gamma) = \lim_{\to X} K_0(X)$ we may restrict to finite complexes. If $X \subset B\Gamma$ is a finite complex, we may appeal to a result by P. Baum and R. Douglas [10] stating that every element x of $K_0(X)$ can be described as a quadruple (M, E, D, f) where:

- M is a smooth compact manifold;
- $E = E^+ \oplus E^-$ is a $\mathbf{Z}_2$-graded smooth vector bundle over M;
- $D : C^\infty(M, E^\pm) \to C^\infty(M, E^\mp)$ is an elliptic differential operator on smooth sections of E;
- $f : M \to X$ is a continuous map.

We form the pull-back:

$$\begin{array}{ccc} \widetilde{M} & \longrightarrow & E\Gamma \\ \downarrow{\scriptstyle p} & & \downarrow \\ M & \xrightarrow[f]{} & B\Gamma \end{array}$$

so that $\widetilde{M}$ is a free, proper Γ-manifold. With $\widetilde{E} = p^*E$ and $\widetilde{D}$ the lift of D to a Γ-invariant elliptic differential operator on sections of $\widetilde{E}$, we can construct, as in Example 4.2.3 (4), a cycle $y \in K_0^\Gamma(\widetilde{M}) = K_0(M)$, such that $f_*(y) = x \in K_0(X)$. Atiyah's L^2-index theorem [2] then says:

$$\tau_*(\mu_0^\Gamma(x)) = \mathrm{Ind}(D)$$

where $\mathrm{Ind}(D)$ is the (usual) Fredholm index of the elliptic operator D.
So $\tau_*(\mu_0^\Gamma(x)) \in \mathbf{Z}$. □

Remarks 6.3.2. (1) In the above proof, the use of the L^2-index theorem is more transparent when one appeals to the unbounded picture for the Baum-Connes assembly map, as explained in [55]. Indeed, in this picture $\mu_0^\Gamma(x)$ is just described by $\widetilde{D}$, acting on the suitable Hilbert C*-module completion of $C^\infty(\widetilde{M}, \widetilde{E})$.

(2) G. Mislin has recently given in [67] a completely different proof of Proposition 6.3.1, based on a clever use of acyclic groups.

Exercise 6.3.3. Let A be a unital C*-algebra, and let $e \in A$ be an idempotent. Set $z = 1 + (e^* - e)^*(e^* - e)$.

- Show that z is invertible and that $ez = ze = ee^*e$.
- Set $p = ee^*z^{-1}$. Show that p is a self-adjoint idempotent ($p^* = p^2 = p$) and that $ep = p$, $pe = e$.

Corollary 6.3.4. *Let Γ be a torsion-free group. If μ_0^Γ is onto, then Conjecture 3 holds, i.e., there is no non-trivial idempotent in $C_r^*\Gamma$.*

Proof. Let e be an idempotent in $C_r^*\Gamma$, and let $[e]$ be its class in $K_0(C_r^*\Gamma)$. Since μ_0^Γ is onto, $\tau_*[e] = \tau(e)$ is an integer. By Exercise 6.3.3, we can find a self-adjoint idempotent $p \in C_r^*\Gamma$ such that $ep = p$ and $pe = e$. In particular, $\tau(e) = \tau(p)$. Now, by Exercise 3.1.11:

$$
\begin{aligned}
\tau(p) &= \tau(p^*p) \geq 0; \\
1 - \tau(p) &= \tau(1-p) = \tau\left((1-p)^*(1-p)\right) \geq 0
\end{aligned}
$$

so that $0 \leq \tau(p) \leq 1$. But $\tau(p)$ is an integer, so that either $\tau(p) = 1$, or $\tau(p) = 0$. Using faithfulness of τ, we get in the first case $p = 1$, hence also $e = 1$, and in the second case $p = 0$, hence also $e = 0$. □

Chapter 7

Some Examples of the Assembly Map

Let Γ denote a countable group. Consider the following diagram:

$$\begin{array}{ccccccc}
 & & RK_i(B\Gamma) & \xrightarrow{\ \iota\ } & RK_i^\Gamma(\underline{E}\Gamma) & \xrightarrow{\ \mu_i^\Gamma\ } & K_i(C_r^*\Gamma) \\
 & \overset{\beta_t}{\nearrow} & \Big\downarrow \mathrm{Ch}_* & \overset{\beta_a}{\cdots\cdots} & \Big\downarrow \mathrm{Ch}_*^\Gamma & \nearrow & \\
H_j(\Gamma,\mathbf{Z}) & \longrightarrow & \bigoplus_{n=0}^{\infty} H_{2n+i}(\Gamma,\mathbf{Q}) & \longrightarrow & \bigoplus_{n=0}^{\infty} H_{2n+i}(\Gamma,F\Gamma) & &
\end{array}$$

where $i = 0, 1$ and $F\Gamma$ is as in Theorem 4.2.12.

The aim of what follows is to construct, for small j's (namely $j = 0, 1, 2$), maps

$$\begin{array}{rcl}
\beta_t : H_j(\Gamma,\mathbf{Z}) & \rightarrow & RK_i(B\Gamma) \\
\beta_a : H_j(\Gamma,\mathbf{Z}) & \rightarrow & K_i(C_r^*\Gamma)
\end{array}$$

such that $\mathrm{Ch}_* \circ (\beta_t \otimes \mathbf{1}) = \mathrm{Id}_{H_j(\Gamma,\mathbf{Q})}$ and the diagram commutes. But we want a direct construction of β_a, so that $\beta_a = \mu_i^\Gamma \circ \iota \circ \beta_t$ is non trivial. This illustrates what the Baum-Connes map does in small homological degree.

The case $j = 0$: Define $\beta_t : \mathbf{Z} \rightarrow RK_0(B\Gamma)$ by mapping 1 to $[i_*]$, the class of the element of $RK_0(B\Gamma)$ corresponding to the inclusion of the base point.

Define $\beta_a : \mathbf{Z} \rightarrow K_0(C_r^*\Gamma)$ by mapping 1 to $[1]$, the K-theory class of the unit. Example 6.1.5 shows that $\mu_0^\Gamma \circ \beta_t = \beta_a$. The canonical trace gives a map $\tau_* : K_0(C_r^*\Gamma) \rightarrow \mathbf{R}$ and $\tau_*([1]) = 1$, that is, β_a is injective.

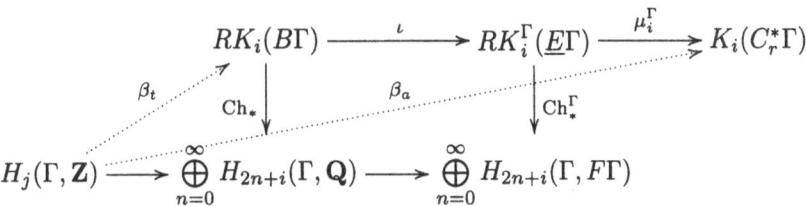

The case $j = 1$: Define $\tilde{\beta}_a : \Gamma \to K_1(C_r^*\Gamma)$ by mapping an element $\gamma \in \Gamma$ to the K-theory class $[\delta_\gamma]$ of the corresponding invertible element $\delta_\gamma \in C_r^*\Gamma$. Since

$$H_1(\Gamma, \mathbf{Z}) = \Gamma^{\mathrm{ab}} = \Gamma/[\Gamma, \Gamma]$$

and $K_1(C_r^*\Gamma)$ is abelian, the group homomorphism $\tilde{\beta}_a$ factors through

$$\beta_a : \Gamma^{\mathrm{ab}} \to K_1(C_r^*\Gamma).$$

To define $\beta_t : \Gamma^{\mathrm{ab}} \to RK_1(B\Gamma)$, first notice that since $\pi_1(B\Gamma) = \Gamma$, we may view an element $\gamma \in \Gamma$ as a pointed continuous map

$$\gamma : S^1 \to B\Gamma,$$

and thus γ induces a map in K-homology,

$$\gamma_* : RK_1(S^1) \to RK_1(B\Gamma).$$

Since $RK_1(S^1) = K_1(S^1) \simeq \mathbf{Z}$ has a single generator that can be described (as in Examples 4.2.3 (2) and (3)) by the class of the cycle (π, D) where for an $f \in C(S^1)$

$$\pi(f) : L^2(S^1) \to L^2(S^1)$$

is the pointwise multiplication by f, and

$$D = -i\frac{d}{dt},$$

with domain $\mathrm{Dom}(D) = \{\xi \in L^2(S^1) | D\xi \in L^2(S^1)\}$, we can map an element $\gamma \in \Gamma$ to the class of the cycle

$$\gamma_*(\pi, D) = (\gamma_*\pi, D)$$

where for X a compact subset of $B\Gamma$ containing $\gamma(S^1)$ and $f \in C(X)$, $\gamma_*\pi(f) = \pi(f \circ \gamma)$ is the pointwise multiplication by $f \circ \gamma$ on $L^2(S^1)$. Hence we define

$$\begin{array}{rcl} \tilde{\beta}_t : \Gamma & \to & RK_1(B\Gamma) \\ \gamma & \mapsto & [(\gamma_*\pi, D)]. \end{array}$$

Proposition 7.1. *The map $\tilde{\beta}_t : \Gamma \to RK_1(B\Gamma)$ as previously defined is a group homomorphism and hence factors through*

$$\beta_t : \Gamma^{\mathrm{ab}} \to RK_1(B\Gamma).$$

Proof. For $\gamma_1, \gamma_2 \in \Gamma$, we have

$$\gamma_1\gamma_2(t) = \begin{cases} \gamma_1(2t) & t \in [0, 1/2] \\ \gamma_2(2t-1) & t \in [1/2, 1] \end{cases}$$

so that $\tilde{\beta}_t(\gamma_1\gamma_2) = [((\gamma_1\gamma_2)_*\pi, D)]$. On the other hand,

$$\begin{aligned}\tilde{\beta}_t(\gamma_1) + \tilde{\beta}_t(\gamma_2) &= [(\gamma_{1*}\pi, D)] + [(\gamma_{2*}\pi, D)] \\ &= [(\gamma_{1*}\pi \oplus \gamma_{2*}\pi, D \oplus D)].\end{aligned}$$

This holds in $K_1(X)$, where X is a compact subset of $B\Gamma$ containing $\gamma_1(S^1) \cup \gamma_2(S^1)$. Now consider the unitary operator

$$\begin{aligned}S : L^2(S^1) \oplus L^2(S^1) &\to L^2(S^1) \\ (\xi_1, \xi_2) &\mapsto \{t \mapsto \begin{cases} \sqrt{2}\xi_1(2t) & t \in [0, 1/2] \\ \sqrt{2}\xi_2(2t-1) & t \in [1/2, 1] \end{cases}\}\end{aligned}$$

whose adjoint is given by

$$\begin{aligned}S^* : L^2(S^1) &\to L^2(S^1) \oplus L^2(S^1) \\ \xi &\mapsto (\{t \mapsto \frac{1}{\sqrt{2}}\xi(\frac{t}{2})\}, \{t \mapsto \frac{1}{\sqrt{2}}\xi(\frac{t+1}{2})\}).\end{aligned}$$

Neglecting questions of domains (we are actually hiding here an analytical difficulty discussed in [13]),

$$S^*DS = 2(D \oplus D)$$

and

$$\begin{aligned}S^*\left((\gamma_1\gamma_2)_*\pi\right)(f)S &= S^*\pi(f \circ \gamma_1\gamma_2)S \\ &= \pi(f \circ \gamma_1) \oplus \pi(f \circ \gamma_2)\end{aligned}$$

for all $f \in C(X)$, this means that

$$s \mapsto (\gamma_{1*}\pi \oplus \gamma_{2*}\pi, (1+s)(D \oplus D)) \quad (s \in [0, 1])$$

gives a homotopy between $S^*((\gamma_1\gamma_2)_*\pi, D)S$ and $(\gamma_{1*}\pi \oplus \gamma_{2*}\pi, D \oplus D)$. Hence

$$\tilde{\beta}_t(\gamma_1\gamma_2) = \tilde{\beta}_t(\gamma_1) + \tilde{\beta}_t(\gamma_2).$$

□

Proposition 7.2. (1) (T. Natsume, [68]) *We have that*

$$\beta_a = \mu_1^\Gamma \circ \iota \circ \beta_t.$$

(2) (G. A. Elliott and T. Natsume, [25], see also [14]) *The homomorphism β_a is rationally injective.*

Proof. (1) (taken from [13]) Clearly, it suffices to prove that

$$\tilde{\beta}_a = \mu_1^\Gamma \circ \iota \circ \tilde{\beta}_t$$

as group homomorphisms $\Gamma \to K_1(C_r^*\Gamma)$. By abuse of notation, we shall write $\tilde{\beta}_t$ instead of $\iota \circ \tilde{\beta}_t$. Fix $\gamma \in \Gamma$, and denote by the same letter γ the unique homomorphism $\mathbf{Z} \to \Gamma$ such that $\gamma(1) = \gamma$. Consider then the diagram

$$\begin{array}{ccccc}
\mathbf{Z} & & \xrightarrow{\quad\gamma\quad} & & \Gamma \\
\Big\downarrow \beta_t & \searrow^{\beta_a} & & \swarrow^{\tilde{\beta}_a} & \Big\downarrow \tilde{\beta}_t \\
 & K_1(C_r^*\mathbf{Z}) & \xrightarrow[\gamma_*]{} & K_1(C_r^*\Gamma) & \\
 & \nearrow_{\mu_1^{\mathbf{Z}}} & & \nwarrow_{\mu_1^\Gamma} & \\
RK_1^{\mathbf{Z}}(\underline{E}\mathbf{Z}) & & \xrightarrow[\gamma_*]{\qquad} & & RK_1^\Gamma(\underline{E}\Gamma)
\end{array}$$

We have $\tilde{\beta}_a \circ \gamma = \gamma_* \circ \beta_a$ trivially, $\tilde{\beta}_a \circ \gamma = \gamma_* \circ \beta_a$ by the very definition of $\tilde{\beta}_t$, and $\gamma_* \circ \mu_1^{\mathbf{Z}} = \mu_1^\Gamma \circ \gamma_*$ by naturality of the assembly map. By diagram chasing, the desired result follows from the analogous result for $\mathbf{Z}$, i.e. $\beta_a = \mu_1^{\mathbf{Z}} \circ \beta_t$. This in turn follows from the proof of Conjecture 1 for $\mathbf{Z}$, given in Example 6.1.6 above.

(2) (sketch) Since every group is the inductive limit of its finitely generated subgroups, we may assume that Γ is finitely generated, so that

$$\Gamma^{\mathrm{ab}} = \mathbf{Z}^n \oplus F,$$

where F is a finite abelian group. Let $\bar{\gamma} \in \mathbf{Z}^n$ be non zero, we want to show that $\beta_a(\bar{\gamma}) \neq 0$. We may assume that $\bar{\gamma}$ is primitive. There exists a map $\mathbf{Z}^n \to \mathbf{Z}$ mapping $\bar{\gamma} \mapsto 1$. Composing it with the quotient map $\Gamma \to \Gamma^{\mathrm{ab}}$, we get a homomorphism

$$\alpha : \Gamma \to \mathbf{Z}$$

which is onto and such that $\alpha(\gamma) = 1$. Hence we can write Γ as a semi-direct product $\Gamma = N \rtimes \mathbf{Z}$, where $N = \ker(\alpha)$. Then the reduced C*-algebra will be given by a crossed product

$$C_r^*\Gamma = C_r^*N \rtimes \mathbf{Z},$$

and we may appeal to the Pimsner-Voiculescu exact sequence (see [74]) to compute $K_*(C_r^*N \rtimes \mathbf{Z})$ (here $i : C_r^*N \to C_r^*\Gamma$ is the inclusion).

$$\begin{array}{ccccc}
K_0(C_r^*N) & \xrightarrow{[\mathrm{Id}]_* - [\alpha]_*} & K_0(C_r^*N) & \xrightarrow{\quad i_* \quad} & K_0(C_r^*\Gamma) \\
\partial_1 \Big\uparrow & & & & \Big\downarrow \partial_0 \\
K_1(C_r^*\Gamma) & \xleftarrow{\quad i_* \quad} & K_1(C_r^*N) & \xleftarrow{[\mathrm{Id}]_* - [\alpha]_*} & K_1(C_r^*N)
\end{array}$$

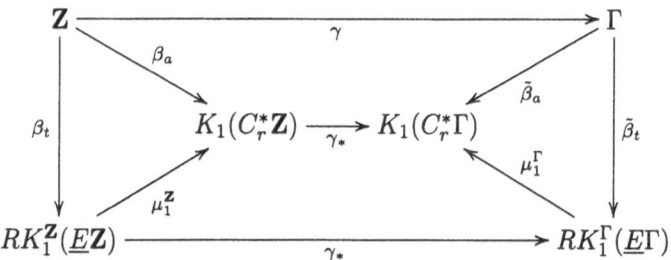

For $\beta_a(\bar{\gamma}) \in K_1(C_r^*\Gamma)$, from the definition of the connecting map ∂_1 in the Pimsner-Voiculescu exact sequence, we have that $\partial_1(\beta_a(\bar{\gamma})) = [1] \in K_0(C_r^*N)$, and $[1] \neq 0$. □

The case $j = 2$:

Proposition 7.3 (see [13]). *There exists maps*

$$\begin{aligned} \beta_t : H_2(\Gamma, \mathbf{Z}) &\rightarrow RK_0(B\Gamma) \\ \beta_a : H_2(\Gamma, \mathbf{Z}) &\rightarrow K_0(C_r^*\Gamma) \end{aligned}$$

such that $\mathrm{Ch}_* \circ (\beta_t \otimes 1) = \mathrm{Id}_{H_2(\Gamma, \mathbf{Q})}$, *and* $\beta_a = \mu_0^\Gamma \circ \iota \circ \beta_t$.

Proof. (sketch) Use a description of $H_2(\Gamma, \mathbf{Z})$ due to B. Zimmermann in [93] by means of pointed continuous maps

$$\Sigma_g \rightarrow B\Gamma,$$

inducing epimorphisms at the level of fundamental groups. Denote by $S(\Sigma_g, B\Gamma)$ the set of these maps. Say that $f_1, f_2 \in S(\Sigma_g, B\Gamma)$ are *equivalent* if, for some orientation preserving homeomorphism h of Σ_g, the maps f_1 and $f_2 \circ h$ are homotopic. We then write $f_1 \sim f_2$. We say that $f_1 \in S(\Sigma_{g_1}, B\Gamma)$ and $f_2 \in S(\Sigma_{g_2}, B\Gamma)$ are *stably equivalent* if they become equivalent after extending them, homotopically trivially, to suitable connected sums: $f_1 \sharp y_0 \sim f_2 \sharp y_0$ (where y_0 is the base point of $B\Gamma$). Denote by $[\Sigma_g]$ the fundamental class of Σ_g. It was proved by B. Zimmermann [93] that the map

$$\begin{aligned} \coprod_{g \geq 1} S(\Sigma_g, B\Gamma) &\rightarrow H_2(\Gamma, \mathbf{Z}) \\ f &\mapsto f_*[\Sigma_g] \end{aligned}$$

induces a bijection between the set of stable equivalence classes on $\coprod_{g\geq 1} S(\Sigma_g, B\Gamma)$ and $H_2(\Gamma, \mathbf{Z})$ (in particular we may represent geometrically addition in $H_2(\Gamma, \mathbf{Z})$ by connected sums of surfaces).

Let $\bar{\partial}_g$ be the Dolbeault operator on Σ_g (associated with an auxiliary complex structure), and let $[\bar{\partial}_g]$ be its class in $K_0(\Sigma_g)$. For $f \in S(\Sigma_g, B\Gamma)$, we set

$$\beta_t(f) = f_*([\bar{\partial}_g] + (g-1)[i_*])$$

(where $[i_*]$ is, as in the case $j = 0$, the element of $RK_0(B\Gamma)$ corresponding to the inclusion of the base point), and

$$\beta_a(f) = (\lambda_\Gamma \circ f)_*(\tilde{\mu}_0^{\Gamma_g}([\bar{\partial}_g] + (g-1)[i_*])).$$

Here $\Gamma_g = \pi_1(\Sigma_g)$, $\tilde{\mu}_0^{\Gamma_g} : K_0(\Sigma_g) \rightarrow K_0(C^*\Gamma_g)$ is the Baum-Connes assembly map at the level of the full C*-algebra $C^*\Gamma_g$ (see [42]), f is viewed as a homomorphism $C^*\Gamma_g \rightarrow C^*\Gamma$, and $\lambda_\Gamma : C^*\Gamma \rightarrow C_r^*\Gamma$ is the left regular representation. It is part of the result that this map is well defined, see [13] for details. □

Remark 7.4. If $B\Gamma$ is a finite 2-complex, then $K_0(B\Gamma) \simeq H_0(\Gamma, \mathbf{Z}) \oplus H_2(\Gamma, \mathbf{Z})$ and $K_1(B\Gamma) \simeq H_1(\Gamma, \mathbf{Z})$, so that we can reformulate Conjecture 1 using β_a instead of μ_i^Γ (this technique has been used to prove Conjecture 1 for torsion-free one-relator groups in [11]).

Remark 7.5. There is a "delocalized" version of β_a and β_t for $j = 0, 1, 2$ (see [63]): $\beta_t : H_j(\Gamma, F\Gamma) \to K_i^\Gamma(\underline{E}\Gamma) \otimes \mathbf{C}$ and $\beta_a : H_j(\Gamma, F\Gamma) \to K_i(C_r^*\Gamma) \otimes \mathbf{C}$.

Chapter 8

A Glimpse into Non-commutative Geometry: Property (RD)

Definition 8.1. A *length function* on a group Γ is a function

$$L : \Gamma \to \mathbf{R}^+$$

such that:

(1) $L(1) = 0$,
(2) $L(\gamma) = L(\gamma^{-1})$ for all $\gamma \in \Gamma$,
(3) $L(\gamma_1\gamma_2) \leq L(\gamma_1) + L(\gamma_2)$ for all $\gamma_1, \gamma_2 \in \Gamma$.

Example 8.2. (a) If Γ is generated by some finite subset S, then the word length $L_S : \Gamma \to \mathbf{N}$ is an example of a length function on Γ (recall that, for $\gamma \in \Gamma$, the *word length* $L_S(\gamma)$ is the minimal length of γ as a word on the alphabet $S \cup S^{-1}$).
(b) Let Γ act by isometries on a metric space (X, d). Pick a point $x_0 \in X$ and define $L(\gamma) = d(\gamma \cdot x_0, x_0)$, this is a length function on Γ (it is an easy exercise to see that this example is universal).

Definition 8.3. For $s > 0$ and $f : \Gamma \to \mathbf{C}$ set

$$\|f\|_{L,s} = \Big(\sum_{\gamma\in\Gamma} |f(\gamma)|^2 (1 + L(\gamma))^{2s}\Big)^{1/2}.$$

This is a *weighted ℓ^2-norm* on Γ.

Definition 8.4 (P. Jolissaint, [40]). Let L be a length function on Γ. We say that Γ has *property (RD)* (standing for *Rapid Decay*) with respect to L, or that it satisfies the *Haagerup inequality* if there exists $C, s > 0$ such that, for each $f \in \mathbf{C}\Gamma$ one has

$$\|\lambda_\Gamma(f)\|_{op} \leq C\|f\|_{L,s},$$

where λ_Γ is the left regular representation of Γ on $\ell^2\Gamma$ and $\|\lambda_\Gamma(f)\|_{op}$ is the norm of f in $C_r^*\Gamma$, i.e. the operator norm of f as left convolutor on $\ell^2\Gamma$.

Example 8.5. (1) If Γ is finitely generated with polynomial growth, then Γ has property (RD) with respect to the word length. Indeed, for a function $f \in \mathbf{C}\Gamma$, we have by Cauchy-Schwarz:

$$\begin{aligned} \|\lambda_\Gamma(f)\|_{op} &\leq \sum_{\gamma\in\Gamma}|f(\gamma)| = \sum_{\gamma\in\Gamma}|f(\gamma)|(1+L(\gamma))^s(1+L(\gamma))^{-s} \\ &\leq \underbrace{\sqrt{\sum_{\gamma\in\Gamma}|f(\gamma)|^2(1+L(\gamma))^{2s}}}_{\|f\|_{L,s}} \underbrace{\sqrt{\sum_{\gamma\in\Gamma}(1+L(\gamma))^{-2s}}}_{C} \end{aligned}$$

and $C < \infty$ for s large because of the polynomial growth.

There is a converse statement, see [40], which is that if Γ is amenable with property (RD), then it has polynomial growth. Indeed, amenability will be used as follows: if $f \in \ell^1\Gamma$, then $|\sum_{\gamma\in\Gamma} f(\gamma)| \leq \|f\|_{op}$ (weak containment of the trivial representation in the regular representation). We apply this to f_k, the characteristic function of B_k the ball of radius k:

$$\begin{aligned} \sharp B_k &= |\sum_{\gamma\in\Gamma} f_k(\gamma)| \leq \|f_k\|_{op} \leq C\|f_k\|_{L,s} \\ &= \sqrt{\sum_{L(\gamma)\leq k} C^2(1+L(\gamma))^{2s}} \leq C(1+k)^s\sqrt{\sharp B_k} \end{aligned}$$

so that $\sharp B_k \leq C^2(1+k)^{2s}$ and thus Γ is of polynomial growth.

(2) Finitely generated free groups have property (RD), (U. Haagerup, [30]; we shall give an unpublished proof by T. Steger, 1996). If Γ is a free group $\mathbf{F}_n$, set

$$S_k = \{\gamma \in \mathbf{F}_n | L_S(\gamma) = k\},$$

the sphere of radius k with respect to the length L. We shall work with ρ, the right regular representation of $\mathbf{F}_n$ on $\ell^2\mathbf{F}_n$. If $g \in \mathbf{C}\mathbf{F}_n$ is such that $\mathrm{supp}(g) \subseteq S_k$ then we will show that

$$\|\rho(g)\|_{op} \leq (k+1)\|g\|_2. \qquad (*)$$

Property (RD) follows from the latter inequality $(*)$ because for $f \in \mathbf{C}\mathbf{F}_n$, write

$f = \sum_{k=0}^{\infty} f_k$ where $f = f_k$ on S_k and 0 elsewhere, then

$$\begin{aligned}\|\rho(f)\|_{op} &\leq \sum_{k=0}^{\infty}\|\rho(f_k)\|_{op} \leq \sum_{k=0}^{\infty}(k+1)\|f_k\|_2 \\ &= \sum_{k=0}^{\infty}(k+1)^2\|f_k\|_2(k+1)^{-1} \\ &\leq \sqrt{\sum_{k=0}^{\infty}(k+1)^4\|f_k\|_2^2}\sqrt{\sum_{k=0}^{\infty}(k+1)^{-2}} = C\|f\|_{L_S,2},\end{aligned}$$

by Cauchy-Schwarz. Now to prove (*), take g such that $\text{supp}(g) \subseteq S_k$. Notice that, for $x, y \in \mathbf{F}_n$ we have

$$\langle \rho(g)\delta_x, \delta_y\rangle = \sum_{\gamma\in\Gamma}(\delta_x * \check{g})(\gamma)\delta_y(\gamma) = (\delta_x * \check{g})(y) = g(y^{-1}x).$$

Fix an end ω of $\mathbf{F}_n$. Let $z = z(x,y)$ be the point of the geodesic $[x,y]$ which is nearest of ω.

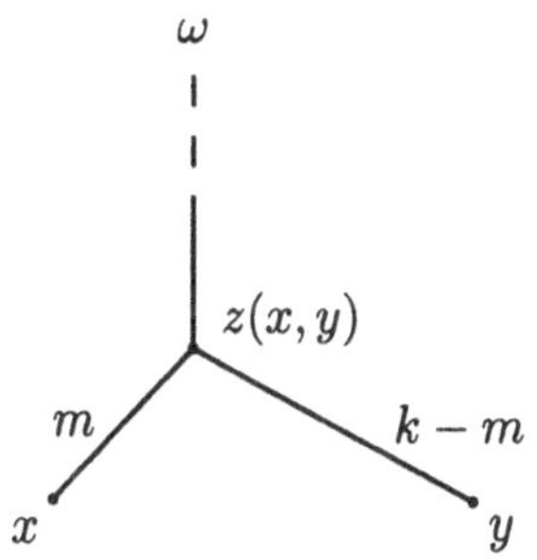

For $0 \leq m \leq k$, define the operator $T^{(m)}$ on $\ell^2\mathbf{F}_n$ by setting

$$< T^{(m)}\delta_x, \delta_y >= \begin{cases} g(y^{-1}x) & \text{if } d(x,z) = m, d(z,y) = k-m \\ 0 & \text{otherwise}\end{cases}$$

Since $g(y^{-1}x) \neq 0$ only if $d(x,y) = k$, we have that $\rho(g) = \sum_{m=0}^{k} T^{(m)}$.

Now let us prove that $\|T^{(m)}\|_{op} \leq \|g\|_2$. Fix m such that $0 \leq m \leq k$. For $z \in \mathbf{F}_n$, denote by $S^\omega(z,m)$ the set of $x \in \mathbf{F}_n$ such that $d(x,z) = m$ and z is between x and ω (this is the sphere of radius m centered at z, without the points x such that the geodesic $[z,x]$ starts towards ω). Let $\mathcal{H}_z \subseteq \ell^2\mathbf{F}_n$ be the linear span of the $\delta_x \in S^\omega(z,m)$ and similarly denote by $\mathcal{K}_z \subseteq \ell^2\mathbf{F}_n$ the linear span of the $\delta_y \in S^\omega(z,k-m)$. We get two decompositions

$$\ell^2\mathbf{F}_n = \bigoplus_{z\in\mathbf{F}_n}\mathcal{H}_z = \bigoplus_{z'\in\mathbf{F}_n}\mathcal{K}_{z'}$$

and thus we can write the operator $T^{(m)}$ as a matrix $(T^{(m)}_{z,z'})_{z,z'\in\mathbf{F}_n}$ where $T^{(m)}_{z,z'}:\mathcal{H}_z\to\mathcal{K}_{z'}$. Now $T^{(m)}$ is block diagonal since $T^{(m)}_{z,z'}=0$ if $z\neq z'$, which means that $\|T^{(m)}\|_{op}=\sup_{z\in\mathbf{F}_n}\|T^{(m)}_{z,z}\|$ and thus it remains to show that $\|T^{(m)}_{z,z}\|\leq\|g\|_2$ for each $z\in\mathbf{F}_n$. We compute

$$\begin{aligned}\|T^{(m)}_{z,z}\| &\leq \|T^{(m)}_{z,z}\|_{\mathrm{H-S}}=\sqrt{\sum_{\substack{x\in S^{\omega}(z,m)\\ y\in S^{\omega}(z,k-m)}} |<T^{(m)}_{z,z}\delta_x,\delta_y>|^2}\\ &= \sqrt{\sum_{\substack{x\in S^{\omega}(z,m)\\ y\in S^{\omega}(z,k-m)}} |g(y^{-1}x)|^2}\leq\|g\|_2.\end{aligned}$$

Here $\|\ \|_{\mathrm{H-S}}$ denotes the Hilbert-Schmidt norm. The last inequality holds because of the fact that for $0\leq m\leq k$ and $z\in\mathbf{F}_n$ fixed, the reduced word $y^{-1}x$ uniquely determines x and y (one can uniquely write $y^{-1}x=\gamma_1\gamma_2$ where the lengths of γ_1 and γ_2 are $k-m$ and m respectively, and so $y=z\gamma_1^{-1}$ while $x=z\gamma_2$).

(3) Hyperbolic groups à la Gromov, see [31].

(4) Co-compact lattices in $SL_3(F)$ where F is a non discrete locally compact field (this is due to J. Ramagge, G. Robertson and T. Steger [76] for non-archimedean fields and to V. Lafforgue [58] for archimedean fields. Recent computations by I. Chatterji [15] allow to add $F=\mathbf{H}$, the Hamilton quaternions, and the exceptional group $E_{6(-26)}$ to that list).

Conjecture 7 (A. Valette). *Property (RD) holds for any discrete group acting isometrically, properly and cocompactly either on a Riemannian symmetric space or on an affine building.*

Remark 8.6. Co-compactness is essential. Indeed, it is easy to see that property (RD) passes to any subgroup with the induced length. Now, $SL_3(\mathbf{Z})$ contains a solvable subgroup with exponential growth:

$$\begin{array}{ccc}\mathbf{Z}^2\rtimes_{\begin{pmatrix}2&1\\1&1\end{pmatrix}}\mathbf{Z} & \hookrightarrow & SL_3(\mathbf{Z})\\ (v,n) & \mapsto & \begin{pmatrix}\begin{pmatrix}2&1\\1&1\end{pmatrix}^n & v\\ 0 & 1\end{pmatrix}\end{array}$$

So $SL_3(\mathbf{Z})$ does not have property (RD). This is one reason why the Baum-Connes Conjecture is still unproved for $SL_3(\mathbf{Z})$.

Proposition 8.7 (Unpublished, E. Leuzinger and C. Pittet). *Any non-uniform lattice in higher rank contains a solvable subgroup with exponential growth (with respect to the word length).*

In the same vein, a result by A. Lubotzky, S. Mozes and M. S. Raghunathan [61] shows that any non-uniform lattice in higher rank contains a cyclic subgroup having exponential growth with respect to the induced length.

Let us explain what property (RD) is good for.

Definition 8.8. Let L be a length function on Γ, and $s \geq 0$. Define

$$H_L^s(\Gamma) = \{f : \Gamma \to \mathbf{C} | \|f\|_{L,s} < \infty\}$$

and

$$H_L^\infty(\Gamma) = \bigcap_{s \geq 0} H_L^s(\Gamma).$$

These are the functions of *rapid decay* on Γ: they decay faster than the inverse of any polynomial in L.

Remark 8.9. If $\Gamma = \mathbf{Z}^k$, then $H_L^s(\Gamma)$ is exactly the s-Sobolev space on the dual group $\hat{\mathbf{Z}}^k \simeq \mathbf{T}^k$. So, in the philosophy of non-commutative geometry, $H_L^\infty(\Gamma)$ is a space of "smooth functions" on the dual of Γ (which is generally a very bad space).

Now we can state the following

Theorem 8.10 (P. Jolissaint, [40]). *If a group Γ has property (RD) with respect to some length function L, then*

(1) *The algebra $H_L^\infty(\Gamma)$ is a dense subalgebra of $C_r^*\Gamma$.*

(2) *The inclusion*

$$H_L^\infty(\Gamma) \hookrightarrow C_r^*\Gamma$$

induces isomorphisms in K-theory.

To prove this theorem we first need to define an unbounded operator D_L on $\ell^2\Gamma$, namely the pointwise multiplication by L, and then an unbounded derivation on $C_r^*\Gamma$ with values in $\mathcal{B}(\ell^2\Gamma)$ by

$$\delta_L : a \mapsto [D_L, a]$$

Exercise 8.11. Prove, by induction over k, that for all $\varphi \in C_r^*\Gamma$, $\xi \in \ell^2\Gamma$ and $\gamma \in \Gamma$ the equality

$$(\delta_L^k(\varphi)\xi)(\gamma) = \sum_{t \in \Gamma} \varphi(t)\xi(t^{-1}\gamma)(L(\gamma) - L(t^{-1}\gamma))^k$$

holds.

Using the exercise, we have by the triangle inequality for L:

$$|(\delta_L^k(\varphi)\xi)(\gamma)| \leq \sum_{t\in\Gamma} |\varphi(t)|\,|\xi(t^{-1}\gamma)|\,L(t)^k = (|\varphi|L^k * |\xi|)(\gamma)$$

and thus

$$\|\delta_L^k(\varphi)\xi\|_2 \leq \|\lambda_\Gamma(|\varphi|L^k)\|_{op}\,\|\xi\|_2.$$

From this we deduce that $\mathbf{C}\Gamma \subseteq \bigcap_{k\geq 0} \mathrm{Dom}(\delta_L^k)$. Moreover,

$$\bigcap_{k\geq 0} \mathrm{Dom}(\delta_L^k) \subseteq H_L^\infty(\Gamma).$$

Indeed, for $\varphi \in \bigcap_{k\geq 0} \mathrm{Dom}(\delta_L^k)$ and $\xi = \delta_1$, again using the exercise above we see that $(\delta_L^k(\varphi)\delta_1)(s) = \varphi(s)L(s)^k$, so that, for all $k \geq 0$, we have that $\varphi L^k \in \ell^2\Gamma$, which means that $\|\varphi\|_{L,k} < \infty$ and thus $\varphi \in H_L^\infty(\Gamma)$.

Now if the group Γ has property (RD), then

$$\bigcap_{k\geq 0} \mathrm{Dom}(\delta_L^k) = H_L^\infty(\Gamma).$$

Indeed, take $\varphi \in H_L^\infty(\Gamma)$, then $\|\varphi\|_{op} \leq C\|\varphi\|_{L,s}$ for some $s > 0$ and

$$\|\delta_L^k(\varphi)\|_{op} \leq \|\varphi L^k\|_{op} \leq C\|\varphi\|_{L,k+s} < \infty,$$

which means that $\varphi \in \bigcap_{k\geq 0} \mathrm{Dom}(\delta_L^k)$.

So we have that

$$\mathbf{C}\Gamma \subseteq H_L^\infty(\Gamma) = \bigcap_{k\geq 0} \mathrm{Dom}(\delta_L^k) \subseteq C_r^*\Gamma.$$

The following result then proves part (1) of Theorem 8.10.

Proposition 8.12 (Ji, [39]). *Let B be a unital Banach algebra and A be a closed unital subalgebra. Let δ be a closed derivation on A taking values in B, with $\delta(1) = 0$. Define*

$$\mathcal{A} = \bigcap_{k\geq 0} \mathrm{Dom}(\delta^k).$$

Then $\mathcal{A}$ is a subalgebra of A, and if it is dense it is spectral, i.e.: if $x \in \mathcal{A}$ is invertible in A, then $x^{-1} \in \mathcal{A}$.

Proof. Here we recall that such a derivation δ of Banach algebras is *closed* if its graph is closed in $A \times B$, which means that for any convergent sequence $a_n \to a$ of A such that $\delta(a_n) \to b$ in B, then $a \in \mathrm{Dom}(\delta)$ (where $\mathrm{Dom}(\delta)$ denotes the domain of δ) and $\delta(a) = b$.

Now that $\mathcal{A}$ is a subalgebra of A follows from

$$\delta^k(ab) = \sum_{m=0}^{k} \binom{k}{m} \delta^m(a)\delta^{k-m}(b).$$

For the statement that $\mathcal{A}$ is spectral if dense, take $x \in \mathcal{A}$ and first assume that $\|x - 1\| < 1$, so that $x^{-1} = \sum_{n=0}^{\infty}(1-x)^n$. Since

$$\delta((1-x)^n) = \sum_{k=1}^{n} (1-x)^{k-1}\delta(1-x)(1-x)^{n-k},$$

we have that

$$\|\delta((1-x)^n)\| \leq n\|1-x\|^{n-1}\|\delta(1-x)\|$$

which means that $\sum_{n=0}^{\infty} \delta((1-x)^n)$ converges in B. As δ is closed, $x^{-1} \in \mathrm{Dom}(\delta)$, and we proceed inductively to show that

$$x^{-1} \in \bigcap_{k \geq 0} \mathrm{Dom}(\delta^k) = \mathcal{A}.$$

Now for $x \in \mathcal{A}$ invertible in A, by density of $\mathcal{A}$ in A, we can find $y \in \mathcal{A}$ such that $\|yx - 1\| < 1$. Then $(yx)^{-1}$ is in $\mathcal{A}$ and hence $x^{-1} = (yx)^{-1}y \in \mathcal{A}$. □

Remark 8.13. The previous proposition shows that for $x \in \mathcal{A}$,

$$\mathrm{Sp}_A(x) = \mathrm{Sp}_{\mathcal{A}}(x),$$

where $\mathrm{Sp}_{\mathcal{A}}(x) = \{\lambda \in \mathbf{C} | (x - \lambda \cdot 1) \text{ is not invertible in } \mathcal{A}\}$. This explains the terminology "spectral".

Actually, we even have that $\mathcal{A}$ is stable under holomorphic functional calculus, that is, for each $x \in \mathcal{A}$ and f holomorphic in a neighborhood of $\mathrm{Sp}_A(x) = \mathrm{Sp}_{\mathcal{A}}(x)$, then $f(x) \in \mathcal{A}$, where

$$f(x) = \frac{1}{2\pi i} \oint_{\gamma} f(z)(z-x)^{-1} dz$$

and γ is a finite system of Jordan curves with disjoint interiors and surrounding $\mathrm{Sp}_A(x)$. Indeed, $f(x)$ is the limit in A of the Riemann sums

$$\sum_{j=1}^{m} f(z_j)(z_j - x)^{-1}\left(\frac{z_j - z_{j-1}}{2\pi i}\right).$$

We apply δ to this Riemann sum and find (using the equality $\delta(a^{-1}) = -a^{-1}\delta(a)a^{-1}$) that

$$\sum_{j=1}^{m} \delta(f(z_j)(z_j - x)^{-1})(\frac{z_j - z_{j-1}}{2\pi i})$$

$$= -\sum_{j=1}^{m} f(z_j)(z_j - x)^{-1}\delta(z_j - x)(z_j - x)^{-1}(\frac{z_j - z_{j-1}}{2\pi i})$$

$$= \sum_{j=1}^{m} f(z_j)(z_j - x)^{-1}\delta(x)(z_j - x)^{-1}(\frac{z_j - z_{j-1}}{2\pi i}),$$

and this converges in B to

$$\frac{1}{2\pi i}\oint_{\gamma} f(z)(z - x)^{-1}\delta(x)(z - x)^{-1}dz.$$

Since the derivation δ is closed, this means that this is $\delta(f(x))$.

The following proposition will complete the proof of the second part of Theorem 8.10.

Proposition 8.14. *Let A be a Banach algebra and $\mathcal{A}$ a dense subalgebra of A. If $\mathcal{A}$ is stable under holomorphic calculus, then the inclusion $\mathcal{A} \subseteq A$ induces isomorphisms in K-theory.*

Proof. **surjectivity**:
$\underline{i=1}$: For $v \in GL_n(A)$, take $\epsilon > 0$ such that for all $u \in M_n(A)$ with $\|u - v\| < \epsilon$, then $u \in GL_n(A)$. By density of $\mathcal{A}$ in A, we may assume that $u \in M_n(\mathcal{A})$, and thus $u \in GL_n(\mathcal{A})$ since $\mathcal{A}$ is spectral. Since u and v are connected in $GL_n(A)$, $[u] = [v]$ in $K_1(A)$.
$\underline{i=0}$: For $e = e^2 \in M_n(A)$, take $x \in M_n(\mathcal{A})$ close enough to e so that the spectrum $\mathrm{Sp}_A(x)$ does not intersect the line $\mathrm{Re}(z) = 1/2$. The spectrum of x has at least two connected components, so that we can define g as to take the value 1 on one connected component of $\mathrm{Sp}_A(x)$ and 0 on the others. Obviously g is holomorphic in a neighborhood of $\mathrm{Sp}_A(x)$ and because $g = g^2$, the element $g(x)$ is an idempotent in A. Since $\mathcal{A}$ is stable under holomorphic calculus, $g(x)$ is in $\mathcal{A}$, and $g(x)$ is close to $g(e) = e$, so we can conclude using Lemma 3.1.8.

injectivity:
$\underline{i=1}$: Let u be an element in $GL_n(\mathcal{A})$ such that $[u] = 0$ in $K_1(A)$. This means that, up to replacing u by $u \oplus 1$ if necessary, u is connected to 1 in $GL_n(A)$. So we find a continuous path $(u_t)_{t\in[0,1]}$ in $GL_n(A)$ such that $u_0 = 1$ and $u_1 = u$. Set

$$\varepsilon = \min\{\|u_t^{-1}\|^{-1} : t \in [0,1]\}.$$

Let $t_0 = 0 < t_1 < \cdots < t_n = 1$ be a subdivision such that $\|u_{t_{i+1}} - u_{t_i}\| < \varepsilon$ ($i = 0, 1, \ldots, n-1$). By density of $\mathcal{A}$ in A, we can find $v_0, v_1, \ldots, v_n \in M_n(A)$ such that $v_0 = 1$, $v_n = u$, $\|v_i - u_{t_i}\| < \varepsilon$ and $\|v_i - v_{i-1}\| < \|v_t^{-1}\|^{-1}$ for $i = 0, 1, \ldots, n-1$ (indeed, the first condition ensures that $v_i \in GL_n(A)$ by Remark 3.2.3 (1)). Consider then the piecewise linear path $v_0 v_1 \ldots v_n$: it is contained in $M_n(\mathcal{A}) \cap GL_n(A)$. Since $\mathcal{A}$ is spectral in A, this path is contained in $GL_n(\mathcal{A})$, so that u is connected to 1 in $GL_n(\mathcal{A})$, and thus $[u] = 0$ in $K_1(\mathcal{A})$.

<u>$i = 0$</u>: Let $[e] - [f] \in K_0(\mathcal{A})$ be such that $[e] = [f]$ in $K_0(A)$ (here $e, f \in M_n(\mathcal{A})$). So there exists $g \in M_k(A)$, idempotent, such that the modules $(e \oplus g)A^{n+k}$ and $(f \oplus g)A^{n+k}$ are isomorphic. By the surjectivity part, we may assume $g \in M_k(\mathcal{A})$. So there exists $m \geq n + k$ and $v \in GL_m(A)$ such that

$$v(e \oplus g \oplus 0)v^{-1} = f \oplus g \oplus 0.$$

Now we can take $u \in GL_m(\mathcal{A})$ close enough to v and thus $u(e \oplus g \oplus 0)u^{-1}$ will be close to $f \oplus g \oplus 0$. So, using Lemma 3.1.8, we conclude that $e \oplus g \oplus 0$ and $f \oplus g \oplus 0$ are conjugate in $GL_m(\mathcal{A})$. So $[e] - [f] = 0$ in $K_0(\mathcal{A})$.

(Notice that we used here Lemma 3.1.8 even if $\mathcal{A}$ is not a Banach algebra. But since $\mathcal{A}$ is spectral, the proof works as well.) □

We close this chapter with an observation recently made by V. Lafforgue:

Proposition 8.15. *Assume that Γ has property (RD) with respect to L. Then $H^s_L(\Gamma)$ is a Banach algebra for s large enough.*

Proof. For $f, g \in \mathbf{C}\Gamma$, $\gamma \in \Gamma$ and $s \geq 1$,we have:

$$\begin{aligned}
&|(f * g)(\gamma)(1 + L(\gamma))^s| \\
&\quad \leq \sum_{\mu \in \Gamma} |f(\mu)|\, |g(\mu^{-1}\gamma)|(1 + L(\mu) + L(\mu^{-1}\gamma))^s \\
&\quad \leq 2^s \left(\sum_{\mu \in \Gamma} |f(\mu)|\, |g(\mu^{-1}\gamma)|(1 + L(\mu))^s \right) \\
&\qquad + 2^s \left(\sum_{\mu \in \Gamma} |f(\mu)|\, |g(\mu^{-1}\gamma)|(1 + L(\mu^{-1}\gamma))^s \right) \\
&\quad = 2^s \left(|f|(1 + L)^s * |g| \right)(\gamma) + 2^s \left(|f| * |g|(1 + L)^s \right)(\gamma)
\end{aligned}$$

Squaring and summing over Γ we get

$$\begin{aligned}
\|f * g\|^2_{L,s} &\leq 2^{2s+1} \left(\|\, |f|(1 + L)^s * |g|\, \|_2^2 + \|\, |f| * |g|(1 + L)^s\, \|_2^2 \right) \\
&\leq 2^{2s+1} \left(\|\, |f|(1 + L)^s\, \|_2^2\, \|\rho(|g|)\|^2_{op} \right) \\
&\quad + 2^{2s+1} \left(\|\lambda(|f|)\|^2_{op} \|\, |g|(1 + L)^s\, \|_2^2 \right) \\
&= 2^{2s+1} \left(\|\lambda(|g|)\|^2_{op} \|f\|^2_{L,s} + \|\lambda(|f|)\|^2_{op} \|g\|^2_{L,s} \right).
\end{aligned}$$

Now, if $C, s > 0$ are such that $\|\lambda(a)\|_{op} \leq C\|a\|_{L,s}$ for every $a \in \mathbf{C}\Gamma$, we get

$$\|f * g\|_{L,s}^2 \leq 2^{2s+2} C^2 \|f\|_{L,s}^2 \|g\|_{L,s}^2$$

which means that $\|\cdot\|_{L,s}$ is a Banach algebra norm. □

Corollary 8.16. *If Γ has property (RD) with respect to L, then, for s large enough, the inclusion $H_L^s \hookrightarrow C_r^*\Gamma$ induces epimorphisms in K-theory.*

Proof. The inclusion $H_L^\infty \hookrightarrow C_r^*\Gamma$ factors through $H_L^s(\Gamma)$ and, by Theorem 8.10, the former induces epimorphisms in K-theory. □

Chapter 9

The Dirac-dual Dirac Method

Definition 9.1. Let A be a Γ-C*-algebra. We say that A is *proper* if there exists a locally compact proper Γ-space X and a Γ-equivariant homomorphism

$$\begin{array}{rcl}\sigma : C_0(X) & \to & \mathcal{B}(A) \\ f & \mapsto & \sigma_f\end{array}$$

such that:

- for all $f \in C_0(X)$ and $a, b \in A$

$$\sigma_f(ab) = a\sigma_f(b) = \sigma_f(a)b$$

  (that is, we view A as a bi-module over itself and require $C_0(X)$ to act on A by endomorphisms of bi-modules);

- if a net $\{f_n\}$ in $C_0(X)$ converges to 1 uniformly on compact subsets of X, then

$$\lim_{n\to\infty} \|\sigma_{f_n}(a) - a\| = 0$$

  for all $a \in A$.

  Note that if A is unital, σ defines a homomorphism

$$\begin{array}{rcl}C_0(X) & \to & Z(A) \\ f & \mapsto & \sigma_f(1),\end{array}$$

where $Z(A)$ denotes the center of A.

Example 9.2. We will say that a vector bundle $E \to X$ is a bundle of algebras if all fibers are algebras (for instance $E = \mathrm{End}(F)$ for $F \to X$ any vector bundle over a base space X). For $E \to X$ a bundle of algebras,

$$A = C_0(X, E)$$

the algebra of continuous sections vanishing at infinity, is a proper algebra, where for each $f \in C_0(X)$, the endomorphism σ_f is the pointwise multiplication of a section by the function f.

Definition 9.3. If $\underline{E}\Gamma$ is the classifying space for proper actions of the group Γ, we will write

$$RKK_*^\Gamma(\underline{E}\Gamma, A) = \varinjlim_X KK_*^\Gamma(C_0(X), A)$$

where X runs along the inductive system of Γ-compact subsets of $\underline{E}\Gamma$.

The following result of J.-L. Tu (see [86]) extends partial results obtained by G. Kasparov [47] and by G. Kasparov and G. Skandalis [51], [52].

Theorem 9.4. *Let Γ be a countable group and let A be a proper Γ-C^*-algebra. Then the Baum-Connes conjecture 5 with coefficients in A holds, i.e. the map*

$$\mu_*^{\Gamma,A} : RKK_*^\Gamma(\underline{E}\Gamma, A) \to K_*(A \rtimes_r \Gamma)$$

is an isomorphism.

Remark 9.5. This theorem allows for a useful strategy to prove Conjecture 1; this strategy has been used in most of the proofs known so far, an exception being the one in [11].

Suppose that, for a given discrete group Γ we can find a proper Γ-C*-algebra A and elements $\alpha \in KK_i^\Gamma(A, \mathbf{C})$ and $\beta \in KK_i^\Gamma(\mathbf{C}, A)$ such that

$$\beta \otimes_A \alpha = 1 \quad \text{in } KK_0^\Gamma(\mathbf{C}, \mathbf{C}),$$

then Conjecture 1 holds. The reason being the following commutative diagram:

$$\begin{array}{ccccc}
RK_*^\Gamma(\underline{E}\Gamma) & \xrightarrow{\otimes_{\mathbf{C}}\beta} & RKK_*^\Gamma(\underline{E}\Gamma, A) & \xrightarrow{\otimes_A \alpha} & RK_*^\Gamma(\underline{E}\Gamma) \\
\downarrow \mu_*^\Gamma & & \simeq \downarrow \mu_*^{\Gamma,A} & & \downarrow \mu_*^\Gamma \\
K_*(C_r^*\Gamma) & \xrightarrow[\otimes_{C_r^*\Gamma} j_\Gamma(\beta)]{} & K_*(A \rtimes_r \Gamma) & \xrightarrow[\otimes_{A \rtimes_r \Gamma} j_\Gamma(\alpha)]{} & K_*(C_r^*\Gamma)
\end{array}$$

and the fact that, using the naturality of j_Γ, composites on the top and the bottom lines are identity. A similar argument shows that Conjecture 5 also follows from $\beta \otimes_A \alpha = 1$ in $KK_0^\Gamma(\mathbf{C}, \mathbf{C})$.

Example 9.6. (a) Let X be a complete Riemannian manifold on which a group Γ acts properly isometrically. Let $T_{\mathbf{C}}^*X$ be the complexified tangent bundle of X and Cliff $T_{\mathbf{C}}^*X$ be the bundle of Clifford algebra. Here we recall that the *Clifford algebra* of a complex vector space E, with respect to a bilinear form g on E is given by the quotient of the tensor algebra

$$T(E) = \mathbf{C} \oplus E \oplus (E \otimes E) \oplus \dots$$

by the two-sided ideal generated by the elements of the form

$$x \otimes x - g(x,x) \cdot 1,$$

for $x \in E$. Now, given the complex bundle $T^*_{\mathbf{C}}X$ over X, we may consider the Clifford bundle Cliff $T^*_{\mathbf{C}}X$, which is the bundle over X whose fibers $\mathrm{Cliff}_x T^*_{\mathbf{C}}X$ are, for each $x \in X$, the Clifford algebra of the complex vector space $T^*_x X$. For an $x \in X$, the multiplication in $\mathrm{Cliff}_x T^*_{\mathbf{C}}X$ is given, for $\xi, \eta \in T^*_x X$ by

$$\xi \cdot \eta + \eta \cdot \xi = 2 \langle \xi | \eta \rangle .$$

Denote by A the sections of Cliff $T^*_{\mathbf{C}}X$ vanishing at infinity, namely

$$A = C_0(X, \mathrm{Cliff}\ T^*_{\mathbf{C}}X).$$

This is a proper Γ-C*-algebra because X is a proper Γ-space. We want to define $\alpha \in KK^\Gamma_0(A, \mathbf{C})$. First take

$$\mathcal{H} = L^2(\bigwedge T^*_{\mathbf{C}}X)$$

the square integrable differential forms on X, with graduation by even and odd forms. Let U be the permutation representation of Γ on $\mathcal{H}$. For f a 1-form on X and $\xi \in \mathcal{H}$, define

$$\pi(f)\xi = \mathrm{ext}(f)\xi + \mathrm{int}(f)\xi = f \wedge \xi + i_f(\xi).$$

Since $\mathrm{ext}(f)\mathrm{int}(f) + \mathrm{int}(f)\mathrm{ext}(f) = \|f\|^2$, π defines an action of A on $\mathcal{H}$. Now $D = d + d^*$ is a self-adjoint operator on $\mathcal{H}$ and defines a bounded operator F by setting

$$F = \frac{D}{\sqrt{1 + D^2}}.$$

Exactly as in Example 4.2.3 (4) this defines an element

$$\alpha = (U, \pi, F) \in KK^\Gamma_0(A, \mathbf{C});$$

this is the *Dirac element.*

Notice that if Γ preserves a Spin^c-structure on X, then there is an alternative description of α explaining the name. A Γ-invariant Spin^c-structure on X is the choice of a Γ-equivariant complex vector bundle S on X, such that

$$\mathrm{Cliff}\ T^*_{\mathbf{C}}X \simeq \mathrm{End}(S)$$

Γ-equivariantly. Suppose the dimension of X is $2m$: the dimension of $\mathrm{Cliff}_x T^*_{\mathbf{C}}X$ is then 2^{2m}, and for each $x \in X$, we want $\mathrm{Cliff}_x T^*_{\mathbf{C}}X$ to be equal to $\mathrm{End}(S_x)$, so S_x is of dimension 2^m. The bundle S is the bundle of *spinors.* We may take $\mathcal{H} = L^2(S)$, and for D the Dirac operator on $\mathcal{H}$. To define it, first choose a connection

$\nabla : C^\infty(S) \to C^\infty(T^*_{\mathbf{C}}X \otimes S)$ and compose it with the Clifford multiplication $\mathrm{Cliff} : C^\infty(T^*_{\mathbf{C}}X \otimes S) \to C^\infty(S)$. Construct then F as above.

(b) Suppose moreover that X is simply connected with non-positive curvature. Denote by d the Riemannian distance and fix $x_0 \in X$. Define $\rho \in C^\infty(X)$ by setting, for each $x \in X$

$$\rho(x) = \sqrt{1 + d(x_0, x)^2}.$$

We want to define $\beta \in KK_0^\Gamma(\mathbf{C}, A)$, where A is defined as in part (a) of this example. Take A as a right C*-module over itself, and

$$U_\gamma \xi = \gamma \cdot \xi$$

for all $\gamma \in \Gamma$ and $\xi \in A$. Denote by $F = \mathrm{Cliff}(d\rho)$ the Clifford multiplication by $d\rho$. Then

$$F^2 - 1 = \|d\rho\|^2 - 1 \quad \text{and} \quad \|d\rho\| = \|\nabla \rho\|.$$

Since $\nabla_x \rho$ is almost a unit tangent vector to the geodesic going from x_0 to x, we have that $\lim_{x\to\infty} \|\nabla_x \rho\| = 1$. Since $F^2 - 1$ is given by the multiplication by an element in A, we see that $F^2 - 1$ is compact in the sense of C*-modules.

Now we also have to check that $[U_\gamma, F]$ is compact for each $\gamma \in \Gamma$. Suppose that we do the same construction with respect to another origin $x_0' \in X$. This defines an operator F' and if we write $\rho'(x) = \sqrt{1 + d(x_0', x)^2}$, then

$$F - F' = \mathrm{Cliff}(d\rho - d\rho')$$

and $(\mathrm{Cliff}(d\rho - d\rho'))^2 = \|d\rho - d\rho'\|^2$. Because of the non-positive curvature of the manifold X, the angle between $\nabla_x \rho$ and $\nabla_x \rho'$ tends to 0 when x is far from x_0 and x_0', that is, $\lim_{x\to\infty} \|d\rho - d\rho'\| = 0$, and hence $F - F'$ is the multiplication by an element of A, so it is compact in the sense of C*-modules. This $\beta = (U, \pi, F)$ is the *dual-Dirac element*.

(c) Assume that the sectional curvature on X is bounded below. The composite element $\beta \otimes_A \alpha \in KK_0^\Gamma(\mathbf{C}, \mathbf{C})$ is described as follows: For $t > 0$, consider $d_t = e^{-t\rho^2/2} d e^{t\rho^2/2}$ and

$$D_t = d_t + d_t^*.$$

Computing, we find that

$$D_t = d + d^* + t \cdot \mathrm{ext}(\rho d\rho) + t \cdot \mathrm{int}(\rho d\rho).$$

We make it bounded by taking

$$F_t = D_t (D_t^2 + 1)^{-1/2}$$

For each $t > 0$, F_t gives the desired element $\beta \otimes_A \alpha \in KK_0^\Gamma(\mathbf{C}, \mathbf{C})$.

For $X = \mathbf{R}^n$, one finds $D_t^2 = \Delta + t^2\rho^2 + t(2k-n)$ on k-forms. Now $\Delta + t^2\rho^2$ is (up to a factor 2) the quantum-mechanics Hamiltonian of the n-dimensional harmonic oscillator. The spectrum of this Hamiltonian is a discrete set of eigenvalues (with finite multiplicity), increasing to infinity: this implies that D_t has compact resolvent, so that F_t is a Fredholm operator.

Theorem 9.7. *Let Γ be a countable group acting properly isometrically on a space X. If X is one of the following spaces:*

$$\mathbf{R}^n\,,\ \mathbf{H}^n(\mathbf{R})\,,\ \mathbf{H}^n(\mathbf{C}),$$

then,

$$\beta \otimes_A \alpha = 1 \in KK_0^\Gamma(\mathbf{C},\mathbf{C}).$$

for α and β the Dirac and dual-Dirac elements and A the algebra of continuous sections vanishing at infinity of the Clifford bundle over X.

For the class of groups appearing in Example (1) Section 2.2, this was basically how Conjecture 1 was proved.

Example 9.8. In the following geometric situations, it is possible to construct a proper Γ-C*-algebra A, a Dirac element $\alpha \in KK_i^\Gamma(A,\mathbf{C})$ and a dual-Dirac element $\beta \in KK_i^\Gamma(\mathbf{C},A)$ (for $i = 0,1$).

(1) Groups acting properly isometrically on a complete Riemannian manifold X with non-positive curvature (see Example 9.6 above).

(2) Groups acting properly on locally finite trees (see [44], and [41] for the construction of A). In that case $\beta \otimes_A \alpha = 1 \in KK_0^\Gamma(\mathbf{C},\mathbf{C})$.

(3) Groups acting properly on locally finite Euclidean buildings (G. Kasparov and G. Skandalis [51]).

(4) Hyperbolic groups (G. Kasparov and G. Skandalis [52]).

(5) a-T-menable groups (N. Higson and G. Kasparov, [34]). Here $\beta \otimes_A \alpha = 1 \in KK_0^\Gamma(\mathbf{C},\mathbf{C})$.

Remarks 9.9. (a) In all cases above it was proved that for X a Γ-compact subset of $\underline{E}\Gamma$, one has

$$\tau_{C_0(X)}(\beta \otimes_A \alpha) = 1 \in KK_0^\Gamma(C_0(X), C_0(X))$$

and this implies the injectivity of the map μ_*^Γ and hence the Novikov conjecture. Indeed, consider the commutative diagram

$$\begin{array}{ccccc}
K_*^\Gamma(X) & \xrightarrow{\otimes_{\mathbf{C}}\beta} & KK_*^\Gamma(C_0(X),A) & \xrightarrow{\otimes_A \alpha} & K_*^\Gamma(X) \\
\downarrow{\scriptstyle \mu_*^\Gamma} & & \downarrow{\scriptstyle \mu^{\Gamma,A}} & & \downarrow{\scriptstyle \mu_*^\Gamma} \\
K_*(C_r^*\Gamma) & \xrightarrow[\otimes_{C_r^*\Gamma} j_\Gamma(\beta)]{} & K_*(A \rtimes_r \Gamma) & \xrightarrow[\otimes_{A\rtimes_r\Gamma} j_\Gamma(\alpha)]{} & K_*(C_r^*\Gamma)
\end{array}$$

If $x \in K_*^\Gamma(X)$ and $\mu_*^\Gamma(x) = 0$, then by Theorem 9.4 (and up to replacing X by a bigger Γ-compact subset of $\underline{E}\Gamma$): $x \otimes_{\mathbf{C}} \beta = 0$ and so,

$$\begin{aligned} 0 &= (x \otimes_{\mathbf{C}} \beta) \otimes_A \alpha = x \otimes_{\mathbf{C}} (\beta \otimes_A \alpha) \\ &= \underbrace{\tau_{C_0(X)}(\beta \otimes_A \alpha)}_{=1} \otimes_{C_0(X)} x = x. \end{aligned}$$

(b) If we analyze what we need to prove surjectivity of μ_*^Γ, we see that it would be enough to have

$$\cdots \otimes_{C_r^*\Gamma} j_\Gamma(\beta \otimes_A \alpha) = I \in \mathrm{End}(K_*(C_r^*\Gamma)). \qquad (*)$$

This would follow from

$$\beta \otimes_A \alpha = 1 \qquad (**)$$

in $KK_0^\Gamma(\mathbf{C}, \mathbf{C})$, as explained in Remark 9.5. But this fails if Γ is infinite with property (T) (observed by A. Connes in 1981, see [44] for a proof). The reason for $(**)$ to fail when Γ has property (T) is the fact that the Γ-representation underlying $\beta \otimes_A \alpha$ is a multiple of the left regular representation, whereas on the other hand, property (T) means that the trivial representation, underlying $1 \in KK_0^\Gamma(\mathbf{C}, \mathbf{C})$, is isolated among the unitary representations. So it will be impossible to realize a homotopy between the trivial and the regular representation.

Relation $(*)$ would also follow from

$$j_\Gamma(\beta \otimes_A \alpha) = 1$$

in $KK_0(C_r^*\Gamma, C_r^*\Gamma)$, but this equality fails to be true if Γ is a lattice in $Sp(n, 1)$ for $n \geq 2$. This negative result due to G. Skandalis [78] ruined the hopes of a general approach of Conjecture 1 based on KK-theory.

Chapter 10

Lafforgue's KK^{Ban} Theory

The basic idea of KK^{Ban}-theory is that Hilbert C*-modules over C*-algebras must be replaced by pairs of Banach modules in duality, over more general Banach algebras.

For a Banach algebra B, a *B-pair* E consists in two Banach spaces $E^<$ and $E^>$, where $E^<$ is a left B-module, $E^>$ a right B-module and there is a B-valued pairing

$$\begin{array}{ccc} E^< \times E^> & \rightarrow & B \\ (x,y) & \mapsto & \langle x|y\rangle_B \end{array}$$

which is B-linear and continuous:

$$\begin{aligned} \langle b\cdot x|y\rangle_B &= b\,\langle x|y\rangle_B \\ \langle x|y\cdot b\rangle_B &= \langle x|y\rangle_B\, b \\ \|\,\langle x|y\rangle_B\,\|_B &\leq \|x\|_{E^<}\|y\|_{E^>}, \end{aligned}$$

for all $b \in B$, $x \in E^<$ and $y \in E^>$. For two given B-pairs E and F, a *morphism of B-pairs* $f : E \to F$ is given by two maps $f^< : F^< \to E^<$ and $f^> : E^> \to F^>$ which are respectively left and right B-modules morphisms and such that, for any $x \in E^>$ and $\eta \in F^<$:

$$\langle \eta | f^>(x)\rangle_B = \langle f^<(\eta)|x\rangle_B\,.$$

We denote by $\mathcal{B}(E,F)$ the space of morphisms of B-pairs from E to F, that we endow with the supremum norm $\|f\| = \max\{\|f^<\|, \|f^>\|\}$. The composition of two morphisms of B-pairs f and g is denoted by fg and is given by the couple $(g\circ f)^< = f^< \circ g^<$ and $(g \circ f)^> = g^> \circ f^>$. The direct sum of two B-pairs E and F is denoted by $E \oplus F$, where $(E\oplus F)^< = E^< \oplus F^<$ and $(E\oplus F)^> = E^> \oplus F^>$, endowed with the ℓ^1 combination of the pairing.

We say that an operator on E is *compact* if it is a norm limit of finite rank operators, and that an operator is of *finite rank* if it is a linear combination of operators of the form

$$\begin{aligned} |\,\eta\rangle\langle\xi\,| : E^< &\to E^< \\ x &\mapsto \langle x\,|\,\eta\rangle_B\,\xi \\ E^> &\to E^> \\ y &\mapsto \eta\,\langle\xi\,|\,y\rangle_B\,, \end{aligned}$$

for $\xi \in E^<$ and $\eta \in E^>$.

Remark 10.1. If B is a C*-algebra and $\mathcal{H}_B = \ell^2(\mathbf{N}) \otimes B$ as in Example 5.1, then the pair $(\mathcal{H}_B, \mathcal{H}_B)$ is a B-pair. More generally, any Hilbert C*-module $\mathcal{E}$ over B gives a B-pair E where $E^< = \mathcal{E}$ with left B action given by $b \cdot x = x \cdot b^*$ $(b \in B, x \in \mathcal{E})$, and $E^> = \mathcal{E}$ with its original B action.

Let A be a Banach algebra. We will call it a Γ-*Banach algebra* if the group Γ acts on A by isometric automorphisms.

Fix a length function L on Γ. Let us explain how V. Lafforgue [57] associates to any pair (A, B) of Γ-Banach algebras, an abelian group

$$KK^{\mathrm{Ban}}_{\Gamma,L}(A, B)$$

which is contravariant in A and covariant in B.

Definition 10.2. The *cycles* in $KK^{\mathrm{Ban}}_{\Gamma,L}(A, B)$ are given by elements of the form

$$\alpha = (U, \pi, F)_E$$

where:

- E is a $\mathbf{Z}_2$-graded B pair ($E^<$ and $E^>$ are both $\mathbf{Z}_2$-graded).
- $U : \Gamma \to \mathcal{B}(E, E)$ is a representation of Γ in the B-pair E by invertible operators in $\mathcal{B}(E, E)$, such that

$$\begin{aligned} \|U^<_\gamma x\|_{E^<} &\le e^{L(\gamma)}\|x\|_{E^<} \\ \|U^>_\gamma y\|_{E^>} &\le e^{L(\gamma)}\|y\|_{E^>} \end{aligned}$$

  for all $\gamma \in \Gamma$, $x \in E^<$ and $y \in E^>$, and preserving the grading on E.
- $\pi : A \to \mathcal{B}(E, E)$ is a representation of A by bounded operators, preserving the grading on E, and *covariant* in the sense that

$$U_\gamma \pi(a) U_{\gamma^{-1}} = \pi(\gamma \cdot a),$$

  for all $a \in A$ and $\gamma \in \Gamma$.

- $F \in \mathcal{B}(E,E)$ is a bounded operator reversing the grading.
- Furthermore, we require the following operators to be compact:

$$\pi(a)(F^2-1)\,,\ \pi(a)[U_\gamma, F]\,,\ [\pi(a), F],$$

for all $a \in A$ and $\gamma \in \Gamma$.

The abelian group $KK^{\mathrm{Ban}}_{L,\Gamma}(A,B)$ is then defined as the quotient of the set of cycles by a suitable homotopy relation. If $L = 0$, we write $KK^{\mathrm{Ban}}_{\Gamma}(A,B)$; if $\Gamma = \{1\}$ we write $KK^{\mathrm{Ban}}(A,B)$.

The functor $KK^{\mathrm{Ban}}_{\Gamma}(A,B)$ is contravariant in A and covariant in B. Functoriality in A is clear. If $\pi : B_1 \to B_2$ is a continuous homomorphism, the image of the B_1-pair E is the B_2-pair $E \otimes_{B_1^+,\pi} B_2$, the quotient of the projective tensor product $E \otimes_\pi B_2$ by the equivalence relation $x(b_1+\lambda)\otimes y \sim x \otimes \pi(b_1+\lambda)y$.

Remarks 10.3. (a) In case $B = \mathbf{C}$, a $\mathbf{C}$-pair E is just the data of two Banach spaces $E^<$, $E^>$ in duality, with duality satisfying

$$|\,\langle x|y\rangle\,| \le \|x\|_{E^<}\|y\|_{E^>}$$

for all $x \in E^<$ and $y \in E^>$.

(b) Lafforgue's theory is compatible with Kasparov's theory, in the sense that if A, B are Γ-C*-algebras, there is a forgetful map

$$\begin{array}{rcl} \iota : KK_0^{\Gamma}(A,B) & \to & KK^{\mathrm{Ban}}_{\Gamma}(A,B) \\ (U,\pi,F) & \mapsto & (U,\pi,F)_{\mathcal{E}} \end{array}$$

where $\mathcal{E}$ is the Hilbert C*-module underlying (U,π,F), seen as B-pair as in Remark 10.1.

(c) The existence of a multiplicative structure in the KK^{Ban}-theory is an open problem. However, there is a natural isomorphism

$$\zeta : K_0(B) \to KK^{\mathrm{Ban}}(\mathbf{C}, B)$$

and a unique homomorphism

$$p : KK^{\mathrm{Ban}}(A,B) \to \mathrm{Hom}(K_i(A), K_i(B)) \qquad (i = 0,1)$$

functorial in A and B which, for $A = \mathbf{C}$ and $i = 0$ coincides with ζ^{-1}. The compatibility with Kasparov's product is as follows: if A and B are C*-algebras, then the following diagram is commutative:

$$\begin{array}{ccc} KK_0(A,B) & \xrightarrow{\ \iota\ } & KK^{\mathrm{Ban}}(A,B) \\ & \searrow^{\otimes_A} & \downarrow p \\ & & \mathrm{Hom}(K_i(A), K_i(B)) \end{array}$$

Definition 10.4. A Banach algebra $\mathcal{A}\Gamma$ is an *unconditional completion* of $\mathbf{C}\Gamma$ if it contains $\mathbf{C}\Gamma$ as a dense subalgebra and if, for $f_1, f_2 \in \mathbf{C}\Gamma$ such that $|f_1(\gamma)| \leq |f_2(\gamma)|$ for all $\gamma \in \Gamma$, we have

$$\|f_1\|_{\mathcal{A}\Gamma} \leq \|f_2\|_{\mathcal{A}\Gamma}.$$

Taking $f_2 = |f_1|$, we notice that $\|f\|_{\mathcal{A}\Gamma} = \||f|\|_{\mathcal{A}\Gamma}$ for all $f \in \mathbf{C}\Gamma$.

Example 10.5. (1) The Banach algebra $\ell^1\Gamma$ is an unconditional completion.

(2) If Γ has property (RD) with respect to a length function L, then for s large enough,

$$H^s_L(\Gamma) = \{f : \Gamma \to \mathbf{C} | \, \|f\|_{L,s} = \|f(1+L)^s\|_2 < \infty\}$$

is a convolution algebra (see 8.15) and an unconditional completion.

(3) Define $\mathcal{A}_{\max}(\Gamma)$ as the completion of $\mathbf{C}\Gamma$ for

$$\|f\|_{\mathcal{A}_{\max}} = \|\lambda_\Gamma(|f|)\|_{op};$$

this is the biggest unconditional completion that embeds in $C^*_r\Gamma$.

(4) The reduced C*-algebra $C^*_r\Gamma$ is in general not an unconditional completion, even for $\Gamma = \mathbf{Z}$ (reason: the supremum norm of trigonometric polynomials does not depend only on the modulus of coefficients).

Definition 10.6. Let B be a Γ-Banach algebra, $\mathcal{A}\Gamma$ an unconditional completion and L a length function on Γ. For an element $b = \sum_{\gamma\in\Gamma} b_\gamma \gamma \in C_c(\Gamma, B)$, consider the element $e^L|b| \in \mathbf{C}\Gamma$, given by $(e^L|b|)_\gamma = e^{L(\gamma)}\|b_\gamma\|_B$ and consider the norm on $C_c(\Gamma, B)$ given by

$$\|b\| = \|e^L|b|\,\|_{\mathcal{A}\Gamma}.$$

We will denote by $\mathcal{A}_L(\Gamma, B)$ the completion of $C_c(\Gamma, B)$ with respect to this norm. This is a *crossed product* of B by Γ, depending both on $\mathcal{A}\Gamma$ and L. In case where $L = 0$, we will just write $\mathcal{A}(\Gamma, B)$.

Exercise 10.7. Show that $\mathcal{A}_L(\Gamma, B)$ is a Banach algebra, with multiplicative structure given by a twisted convolution.

There is a *descent homomorphism*

$$j_{\mathcal{A}} : KK^{\mathrm{Ban}}_{\Gamma,L}(A, B) \to KK^{\mathrm{Ban}}(\mathcal{A}_L(\Gamma, A), \mathcal{A}(\Gamma, B))$$

which is compatible with the descent homomorphism in KK-theory.

If $\mathcal{A}\Gamma$ is an unconditional completion, one may define an *assembly map*

$$\mu_0^{\mathcal{A}\Gamma} : RK_0^\Gamma(\underline{E}\Gamma) \to K_0(\mathcal{A}\Gamma).$$

Indeed, let X be a Γ-compact subset in $\underline{E}\Gamma$ and let α be a cycle in $K_0^\Gamma(X) = KK_0^\Gamma(C_0(X), \mathbf{C})$. The forgetful map

$$\iota : KK_0^\Gamma(C_0(X), \mathbf{C}) \to KK_\Gamma^{\mathrm{Ban}}(C_0(X), \mathbf{C})$$

composed with the descent homomorphism

$$j_{\mathcal{A}} : KK_\Gamma^{\mathrm{Ban}}(C_0(X), \mathbf{C}) \to KK^{\mathrm{Ban}}(\mathcal{A}(\Gamma, C_0(X)), \mathcal{A}\Gamma)$$

and then with

$$p : KK^{\mathrm{Ban}}(\mathcal{A}(\Gamma, C_0(X)), \mathcal{A}\Gamma) \to \mathrm{Hom}(K_0(\mathcal{A}(\Gamma, C_0(X))), K_0(\mathcal{A}\Gamma))$$

maps α to $p \circ j_{\mathcal{A}} \circ \iota(\alpha) \in \mathrm{Hom}(K_0(\mathcal{A}(\Gamma, C_0(X))), K_0(\mathcal{A}\Gamma))$. Now consider an idempotent $e \in C_c(\Gamma \times X)$ built as in Section 6.2, namely

$$e(\gamma, x) = \sqrt{h(x)h(\gamma^{-1}x)}$$

(where $h \in C_c(X)$ is positive and such that $\sum_{\gamma\in\Gamma} h(\gamma x) = 1$ for all $x \in X$). This defines an element $[e] \in K_0(\mathcal{A}(\Gamma, C_0(X)))$ and we define

$$\begin{array}{rcl} \mu_*^{\mathcal{A}\Gamma} : RK_*^\Gamma(\underline{E}\Gamma) & \to & K_*(\mathcal{A}\Gamma) \\ \alpha & \mapsto & (p \circ j_{\mathcal{A}} \circ \iota(\alpha))([e]). \end{array}$$

Similarly, if A is a Γ-C*-algebra there is an assembly map with coefficients in A,

$$\mu_i^{\mathcal{A}\Gamma, A} : RK_i^\Gamma(\underline{E}\Gamma, A) \to K_i(\mathcal{A}(\Gamma, A)).$$

It is compatible with the Baum-Connes assembly map $\mu_i^{\Gamma,A}$ in the sense that, if $\mathcal{A}(\Gamma, A)$ embeds continuously in $A \rtimes_r \Gamma$, denoting by $i : \mathcal{A}(\Gamma, A) \hookrightarrow A \rtimes_r \Gamma$ the inclusion, we have $i_* \circ \mu_i^{\mathcal{A}\Gamma, A} = \mu_i^{\Gamma,A}$ (see [59]).

Theorem 10.8. *If A is a proper Γ-C^*-algebra, then $\mu_i^{\mathcal{A}\Gamma,A}$ is an isomorphism for every unconditional completion $\mathcal{A}\Gamma$.*

The following result ensures *surjectivity* of $\mu_i^{\mathcal{A}\Gamma}$.

Lemma 10.9. *Let Γ be a group for which there exists a proper Γ-C^*-algebra A, a Dirac element $\alpha \in KK_0^\Gamma(A, \mathbf{C})$ and a dual-Dirac element $\beta \in KK_0^\Gamma(\mathbf{C}, A)$. Assume that there exists a sequence of length functions $(L_n)_{n\geq 0}$, decreasing to 0 and such that*

$$\iota(\beta \otimes_A \alpha) = \iota(1)$$

in $KK_{\Gamma,L_n}^{\mathrm{Ban}}(\mathbf{C}, \mathbf{C})$, for all $n \geq 0$. Then the assembly map

$$\mu_i^{\mathcal{A}\Gamma} : RK_i^\Gamma(\underline{E}\Gamma) \to K_i(\mathcal{A}\Gamma)$$

is onto.

Proof. First notice that

$$\mathcal{A}\Gamma = \mathcal{A}(\Gamma, \mathbf{C}) = \lim_{n\to\infty} \mathcal{A}_{L_n}(\Gamma, \mathbf{C}).$$

By assumption on α and β,

$$p \circ j_{\mathcal{A}} \circ \iota(\beta \otimes_A \alpha) = Id$$

in $\mathrm{End}(K_i(\mathcal{A}_{L_n}(\Gamma, \mathbf{C}))$, and therefore also in $\mathrm{End}(K_i(\mathcal{A}\Gamma))$. So

$$p \circ j_{\mathcal{A}} \circ \iota(\alpha) : K_i(\mathcal{A}(\Gamma, A)) \to K_i(\mathcal{A}\Gamma)$$

is onto. Now we have a commutative diagram

$$\begin{array}{ccc} RKK_i^{\Gamma}(\underline{E}\Gamma, A) & \xrightarrow{\otimes_A \alpha} & RK_i^{\Gamma}(\underline{E}\Gamma) \\ {\scriptstyle \mu_i^{\mathcal{A}\Gamma, A}}\downarrow{\scriptstyle \simeq} & & \downarrow{\scriptstyle \mu_i^{\mathcal{A}\Gamma}} \\ K_i(\mathcal{A}(\Gamma, A)) & \xrightarrow{p \circ j_{\mathcal{A}} \circ \iota(\alpha)} & K_i(\mathcal{A}\Gamma) \end{array}$$

and thus $\mu_i^{\mathcal{A}\Gamma}$ is onto. □

Proposition 10.10. *Let Γ be a group acting properly isometrically either on a simply connected complete Riemannian manifold (M, d) with non-positive curvature, bounded from below, or on a Euclidean building. Then the assumptions on the preceding lemma are satisfied, where we take*

$$L_n(\gamma) = d(\gamma \cdot x_0, x_0)/n$$

for the sequence of length functions decreasing to 0, and $x_0 \in M$ is a fixed base point.

In the Riemannian case, the algebra A and the elements α and β from Lemma 10.9 are those described in Example 9.6.

See Skandalis' recent Bourbaki Seminar [80] for a nice exposition of the proof in the building case.

Therefore, using Lemma 10.9, we see that for every unconditional completion $\mathcal{A}\Gamma$ of a group Γ as in Proposition 10.10, the map $\mu_i^{\mathcal{A}\Gamma}$ is onto, for $i = 0, 1$. Since the assembly maps μ_i^{Γ} and $\mu_i^{\mathcal{A}\Gamma}$ are compatible, it remains to determine when one can find an embedding

$$\mathcal{A}\Gamma \hookrightarrow C_r^*\Gamma$$

that induces epimorphisms in K-theory. If the group Γ has property (RD), this is true (see 8.16), taking $\mathcal{A}\Gamma = H_L^s(\Gamma)$, for s large enough, as an unconditional completion. This completes a rough sketch of the proof of the following remarkable Theorem.

Theorem 10.11 (V. Lafforgue, [57]). *If Γ is a co-compact lattice in a rank one simple Lie group or in $SL_3(\mathbf{R})$ or $SL_3(\mathbf{C})$, then the Baum-Connes conjecture holds for Γ.*

To conclude, let us mention the following conjecture, due to J.-B. Bost (see [80]): for any group Γ, the map

$$\mu_i^{\ell^1\Gamma} : RK_i(\underline{E}\Gamma) \to K_i(\ell^1\Gamma)$$

is an isomorphism. For the groups of geometric interest appearing in Proposition 10.10, this is indeed the case: surjectivity follows from Proposition 10.10; by the compatibility of $\mu_i^{\ell^1\Gamma}$ and μ_i^{Γ}, injectivity of $\mu_i^{\ell^1\Gamma}$ follows from injectivity of μ_i^{Γ}, which was proved by Kasparov [45] in the Riemannian case, and by Kasparov-Skandalis [51] in the building case.

Note that, confronting the Bost conjecture with the Baum-Connes conjecture, one is led to the following conjectural statement, of a purely analytical nature: for any group Γ, the inclusion $\ell^1\Gamma \hookrightarrow C_r^*\Gamma$ induces isomorphisms in K-theory.

Appendix
On the Classifying Space for Proper Actions – Notes for Analysts

by Guido Mislin

The purpose of this Appendix is to explain why the G-CW-model $\underline{E}G$ and the metric model $\underline{\mathcal{E}}G$ of the classifying space of proper actions of a discrete group G are G-homotopy equivalent. Moreover, we recall some basic facts concerning G-CW-complexes and spectra in the topologist's sense.

A.1 The topologist's model

Let G be an arbitrary discrete group. In this section, we will work in the category of G-CW-complexes. The precise definition of a G-CW-complex is recalled in Section A.3; it is a space obtained by gluing together G-cells of the form $G/H \times \sigma$ in quite an analogous way as one glues together cells σ to obtain an ordinary CW-complex. The *usual* topology on a (G)-CW-complex X is often referred to as the *weak* topology; it is characterized by the property that a subset of X is closed if and only if its intersection with each finite subcomplex of X is closed. It is important to keep this in mind when one deals with products (or joins) of CW-complexes: the resulting spaces are in general going to be CW-complexes only if one equips them with this weak topology; we always assume this done if we speak about the *product* or *join* of CW-complexes. By forming products, joins, increasing unions etc. of G-CW-complexes one thus obtains new G-CW-complexes. Note that a zero-dimensional G-CW-complex is the same as a G-set, with its discrete topology.

A G-CW-complex is called *proper* if all point stabilizers are finite (equivalently, if all its G-cells are of the form $G/H \times \sigma$ with H a *finite* subgroup of G). There exists a proper G-CW-complex denoted by "$\underline{E}G$" which plays the same role in the homotopy category of proper G-CW-complexes as a point does in the homotopy category of all spaces:

- For any proper G-CW-complex X there is a unique G-homotopy class of G-maps $\epsilon_X : X \to \underline{E}G$.

Clearly, $\underline{E}G$ is characterized uniquely up to G-homotopy by this property. On the other hand, a simple application of obstruction theory in the homotopy category of proper G-CW-complexes (cf. [64, Chap. I, Sec. 5]) implies that a proper G-CW-complex Z is G-homotopy equivalent to $\underline{E}G$ if and only if the fixed point subcomplexes Z^H are contractible for every finite $H < G$ (note that the fixed point subcomplexes Z^S are empty for infinite $S < G$, since Z is assumed to be a proper G-CW-complex); in particular, $\underline{E}G$ is equivariantly contractible if and only if G is finite. The above characterization of $\underline{E}G$ leads to the following simple construction. Let M be the zero-dimensional G-CW-complex given by the disjoint union of all (left) cosets G/H, H finite, with usual left G-action. Let $M(n)$ denote the n-fold join of M; it is an n-dimensional proper G-CW-complex. There are obvious inclusions $M(n) \to M(n+1)$ and we put

$$\underline{E}G := \bigcup_{n \in \mathbf{N}} M(n).$$

It is easy to verify that this proper G-CW-complex satisfies indeed $(\underline{E}G)^H \simeq \{*\}$ for all finite subgroups $H < G$ (cf. Section A.3). In case G is torsion-free, one has $\underline{E}G = EG = G * G * G \cdots$, the classical CW-model of the universal principal G-space. In general, the space $\underline{E}G$ is referred to as the *classifying space for proper actions* (see [9, Appendix 3] concerning this terminology). Note also that $\underline{E}G$ is **not** metrizable: it contains as a subcomplex the infinite join $\{e\} * \{e\} * \ldots$, which is not locally finite; however, simplicial spaces are metrizable (in their weak topology) if and only if they are locally finite (cf. [81, Thm. 8 of Chap. 3, Sec. 2]).

The following is another description of a standard model for $\underline{E}G$. Let $\mathrm{Fin}(G)$ denote the G-poset of finite non-empty subsets of G, the partial order being the obvious one and the G-action given by left translation. It follows that the geometric realization (with its *weak* topology) $|\mathrm{Fin}(G)|$ of $\mathrm{Fin}(G)$ is a G-CW-complex of type $\underline{E}G$: it is a proper G-CW-complex with contractible H-fixed subcomplexes for all finite $H < G$. Note also that a group homomorphism $\phi : G_1 \to G_2$ induces ϕ-equivariant maps

$$\mathrm{Fin}(G_1) \to \mathrm{Fin}(G_2), \quad \text{and} \quad |\mathrm{Fin}(G_1)| \to |\mathrm{Fin}(G_2)|.$$

Up to G-homotopy there is therefore a well-defined ϕ-equivariant map

$$\underline{E}\phi : \underline{E}G_1 \to \underline{E}G_2,$$

and we can view $\underline{E}(?)$ as a functor on the appropriate category.

A.2 The analyst's model

Following [9, Sec. 2] one defines a G-space $\underline{\mathcal{E}}G$ by putting

$$\underline{\mathcal{E}}G = \{f : G \to [0,1] \mid f \text{ has finite support and } \sum_{x \in G} f(x) = 1\},$$

with obvious (left) G-action $(g \cdot f(x) = f(g^{-1}x), g, x \in G)$. The topology of $\underline{\mathcal{E}}G$ is given by the metric

$$d(f_1, f_2) = \sup_{x \in G} |f_1(x) - f_2(x)|.$$

This is easily seen to be the same topology as the subspace topology induced by the natural embedding

$$\underline{\mathcal{E}}G \subset \prod_G [0,1],$$

where this time the product is equipped with the product topology! Clearly G acts by isometries on $\underline{\mathcal{E}}G$, with finite point stabilizers ($\underline{\mathcal{E}}G$ is a proper G-space in the sense of [9], i.e., it is the union of open subspaces G-homeomorphic to G-spaces of the form $G \times_H Y$, where $H < G$ is finite and Y is a H-space). For each $x \in G$ there is a projection

$$\pi_x : \underline{\mathcal{E}}G \to [0,1], \quad f \mapsto f(x).$$

These *coordinates* $\{\pi_x(f)\}_{x \in G}$ are called the barycentric coordinates of $f \in \underline{\mathcal{E}}G$. Note that a map $\alpha : X \to \underline{\mathcal{E}}G$ is continuous if and only if all coordinate functions $\pi_x \circ \alpha$ are continuous. Since $\sum_{x \in G} \pi_x(f) = 1$, the functions $\{\pi_x\}$ define a (in general not locally finite) partition of unity of $\underline{\mathcal{E}}G$. Since for $g, x \in G$ one has $\pi_x(g \cdot f) = f(g^{-1}x) = \pi_{g^{-1}x}(f)$, it follows that a map $\theta : \underline{\mathcal{E}}G \to \underline{\mathcal{E}}G$ is equivariant if and only if for all $g, x \in G$ one has $\pi_x(\theta(g \cdot f)) = \pi_{g^{-1}x}(\theta(f))$.

We can think of $\underline{\mathcal{E}}G$ as a simplicial complex, with vertices corresponding to the elements of G and n-simplices corresponding to the elements of G^{n+1}. The metric on $\underline{\mathcal{E}}G$ restricts on each simplex σ to the standard metric which, expressed in barycentric coordinates (s_x) resp. (t_x) of two points s and t in σ, is $d(s,t) = \max_x |s_x - t_x|$.

The barycentric subdivision of $\underline{\mathcal{E}}G$ is, as a set, just $|\operatorname{Fin}(G)|$, the geometric realization of the poset of finite non-empty subsets of G. Thus, if we write $|F(G)|_m$ for $|F(G)|$ with the metric topology coming from the metric on each simplex, then $\underline{\mathcal{E}}G = |\operatorname{Fin}(G)|_m$. Note that the topologies of $|\operatorname{Fin}(G)|$ and $|\operatorname{Fin}(G)|_m$ agree on finite subcomplexes so that the identity map

$$\iota : |\operatorname{Fin}(G)| \to |\operatorname{Fin}(G)|_m$$

is continuous. (For a general discussion of metric topologies on simplicial spaces see [22]).

Theorem A.2.1. *The identity map* $\iota : |\operatorname{Fin}(G)| \to \underline{\mathcal{E}}G = |\operatorname{Fin}(G)|_m$ *is a* G-*homotopy equivalence.*

Proof. We will describe a (continuous) G-homotopy inverse to the identity map, following the non-equivariant proof in [21, Appendix A.2]. Maps are described by their effect on barycentric coordinates. Using the functions

$$\mu : |\operatorname{Fin}(G)|_m \to [0,1], \quad f \mapsto \max_{x \in G}(\pi_x(f))$$

and for each $x \in G$

$$\tau_x : |\operatorname{Fin}(G)|_m \to [0,1], \quad f \mapsto \max(0, 2 \cdot \pi_x(f) - \mu(f))$$

one defines a new *locally finite* partition of unity

$$\{\rho_x : |\operatorname{Fin}(G)|_m \to [0,1] \mid x \in G\}$$

by putting $\rho_x(f) = \tau_x(f)/\sum_{x\in G}(\tau_x(f))$. This is then used to define a continuous G-map

$$\kappa : |\operatorname{Fin}(G)|_m \to |\operatorname{Fin}(G)|, \quad \pi_x(\kappa(f)) = \rho_x(f),$$

which is G-homotopy inverse to ι. Indeed, that the compositions $\kappa \circ \iota$ and $\iota \circ \kappa$ are G-homotopic to the identity can be seen from the deformation $\theta_t : f \mapsto f_t$, given by

$$\pi_x(f_t) = t \cdot \rho_x(f) + (1-t) \cdot \pi_x(f).$$

The G-equivariance of θ_t follows from the fact that $\rho_x(g \cdot f) = \rho_{g^{-1}x}(f)$, which implies

$$\begin{aligned} \pi_x(\theta_t(g \cdot f)) &= t \cdot \rho_x(g \cdot f) + (1-t)\pi_x(g \cdot f) = \\ &= t \cdot \rho_{g^{-1}x}(f) + (1-t)\pi_{g^{-1}x}(f) \\ &= \pi_{g^{-1}x}(\theta_t(f)) = \pi_x(g \cdot \theta_t(f)) \end{aligned}$$

thus $\theta_t(g \cdot f) = g \cdot \theta_t(f)$. □

A.3 On G-CW-complexes

Let G be a discrete group. A *G-CW-complex* consists of a Hausdorff space X together with an action of G by self-homeomorphisms and a filtration $X^0 \subseteq X^1 \subseteq X^2 \subseteq \cdots \subseteq X$ by G-subspaces such that the following axioms hold:

1. Each X^n is closed in X.
2. $\bigcup_{n \in \mathbf{N}} X^n = X$.
3. X^0 is a discrete subspace of X.

4. For each $n \geq 1$ there is a discrete G-space Δ_n together with G-maps $f : S^{n-1} \times \Delta_n \to X^{n-1}$ and $\widehat{f} : B^n \times \Delta_n \to X^n$ such that the following diagram is a push-out diagram:

$$\begin{array}{ccc} S^{n-1} \times \Delta_n & \xrightarrow{f} & X^{n-1} \\ \downarrow & & \downarrow \\ B^n \times \Delta_n & \xrightarrow{\widehat{f}} & X^n. \end{array}$$

5. A subspace Y of X is closed if and only if $Y \cap X^n$ is closed for each $n \geq 0$.

Here, we write S^{n-1} and B^n for the standard unit sphere and unit ball in Euclidean n-space, and the vertical maps in the diagram are inclusions. It is useful to adopt the conventions $X^{-1} = \emptyset$ and $\Delta_0 = X^0$. Then for all $n \geq 0$, the nth *cellular* chain group $C_n(X)$ can be defined to be the nth singular homology of the pair X^n, X^{n-1}, and it follows from the Eilenberg-Steenrod Axioms that this is isomorphic as a G-module to the permutation module $\mathbf{Z}\Delta_n$ determined by Δ_n. In effect, $C_n(X)$ is the free abelian group on the G-set of n-cells in X.

A G-CW-complex X is said to be finite dimensional if and only if $X^n = X$ for some n, in which case the dimension is the least $n \geq -1$ for which this happens. For the finite dimensional case, Axiom 5 is redundant.

The most important construction we shall need is that of *join*. For convenience we include a definition here. Further discussion can be found in [20, Chap. I, Sec. 6]. For each $n \geq 0$ let σ^n denote the standard n-simplex in $\mathbf{R}^{n+1}$:

$$\sigma^n = \left\{(t_0, \ldots, t_n) \,\middle|\, \sum t_i = 1,\ t_i \geq 0\right\}.$$

Given CW-complexes $X_0, \ldots, X_n$, the join $X_0 * \cdots * X_n$ is defined to be the identification space

$$(\sigma^n \times X_0 \times \cdots \times X_n)/\sim$$

where

$$(t_0, \ldots, t_n; x_0, \ldots, x_n) \sim (t'_0, \ldots, t'_n; x'_0, \ldots, x'_n)$$

if and only if for each i, either $(t_i, x_i) = (t'_i, x'_i)$ or $t_i = t'_i = 0$. Note that if X_i has dimension d_i then the join has dimension $n + \sum d_i$. Moreover, if G is a group and the X_i are G-CW-complexes then the join inherits a G-CW-structure.

The following are some elementary remarks. For spheres one has $S^n * S^m \cong S^{n+m+1}$. The join increases connectivity: if X, Y, Z are non-empty, then $X * Y$ is connected, $X * Y * Z$ is simply connected etc.; an infinite join of non-empty spaces is contractible. It follows that if $\{X_i | i \in \mathbf{N}\}$ are G-CW-complexes and $H < G$ a subgroup for which X_i^H is non-empty for every i, then $*_i X_i^H = (*X_i)^H$ is contractible; this explains why $\underline{E}G^H$ is contractible for every finite subgroup $H < G$.

A.4 Spectra

The *spectra* in the topologist's sense are convenient objects to describe (co)homology theories in the category of CW-complexes. We will give a brief outline of some definitions; for details, the reader is (for instance) referred to Switzer's book [84]. A *CW-spectrum* is a collection of pointed CW-complexes $\mathbf{S} = \{S_i \mid i \in \mathbf{N}\}$ together with maps $\sigma_i : \Sigma S_i \to S_{i+1}$. The *homotopy groups* $\pi_k(\mathbf{S})$ for $k \in \mathbf{Z}$ are defined by

$$\pi_k(\mathbf{S}) = \operatorname{dirlim}_i \pi_{k+i}(S_i),$$

with the direct limit being taken using the maps σ_i:

$$\pi_{k+i}(S_i) = [S^{k+i}, S_i]_\bullet \xrightarrow{\text{susp}} [S^{k+i+1}, \Sigma S_i]_\bullet \xrightarrow{\sigma_{i\sharp}} [S^{k+i+1}, S_{i+1}]_\bullet = \pi_{k+i+1}(S_{i+1}).$$

Notice that the groups $\pi_k(\mathbf{S})$ can be non-zero for negative values of k. The smash product of a pointed space Y with a spectrum $\mathbf{S}$ yields a new spectrum $Y \wedge \mathbf{S}$ in an obvious way. One also defines homotopy classes of maps between spectra $\mathbf{S}$ and $\mathbf{T}$, $[\mathbf{S}, \mathbf{T}]$, which form naturally an abelian group such that suspension yields an isomorphism (cf. [84])

$$[\mathbf{S}, \mathbf{T}] \cong [S^1 \wedge \mathbf{S}, S^1 \wedge \mathbf{T}].$$

The simplest example of a CW-spectrum is the *sphere* spectrum $\mathbf{Sph}$, which has $S_i = S^i$ and $\sigma_i = Id$ (up to a homeomorphism). The resulting groups

$$\pi_k(\mathbf{Sph}) = \pi_k^{st}(S^0)$$

are the *stable* homotopy groups of the zero sphere. More generally, for any pointed space Y one has

$$\pi_k(Y \wedge \mathbf{Sph}) = \pi_k^{st}(Y).$$

Stable homotopy theory is an example of a *generalized* homology theory. Every spectrum $\mathbf{S}$ gives rise to a generalized (co)homology theory on CW-complexes, and conversely every generalized (co)homology theory $\mathbf{h}$ on CW-complexes (satisfying the usual axioms, including: *homotopy invariance, Mayer-Vietoris axiom and the general disjoint union axiom*) is representable by a spectrum in the sense that

$$\mathbf{h}_i(X) = \pi_i(X_+ \wedge \mathbf{S}) = [S^i \wedge \mathbf{Sph}, X_+ \wedge \mathbf{S}],$$

and

$$\mathbf{h}^i(X) = [X_+ \wedge \mathbf{Sph}, S^i \wedge \mathbf{S}].$$

The notation X_+ stands for X with a disjoint base-point added. If one writes X as a union of subcomplexes X_α such that $X = \operatorname{dirlim}_\alpha X_\alpha$ (e.g., X written as union of its finite subcomplexes, or X written as union over its skeleta), then one always has

$$\mathbf{h}_i(X) = \operatorname{dirlim}_\alpha \mathbf{h}_i(X_\alpha).$$

This can be paraphrased by saying that generalized homology theories (given by spectra) always have *compact supports.* If one wishes to extend a generalized (co)homology theory to the category of *all* spaces, one can do this by defining $\mathbf{h}(X)$ as $\mathbf{h}(|SX|)$, where $|SX|$ denotes the geometric realization of the singular complex of X. The resulting (co)homology theory then obviously satisfies the strong form of the homotopy axiom, turning weak homotopy equivalences into isomorphisms.

Every spectrum is equivalent to an Ω-spectrum, that is, a spectrum $\mathbf{S}$ for which the adjoints $S_i \to \Omega S_{i+1}$ of the maps σ_i are homotopy equivalences. In case one uses an Ω-spectrum to represent $\mathbf{h}$, the cohomology groups can be expressed as *ordinary* homotopy groups:

$$\mathbf{h}^i(X) = [X_+, S_i]_\bullet = [X, S_i].$$

The most famous examples of Ω-spectra are the Eilenberg–Mac Lane spectrum $\mathbf{H}$ which represents ordinary (co)homology, and the Bott spectrum $\mathbf{BU}$ representing (complex) K-(co)homology. They are given by $\mathbf{H} = \{K(i, \mathbf{Z})\}$, with $\Sigma K(i, \mathbf{Z}) \to K(i+1, \mathbf{Z})$ being the adjoint of the natural equivalence $K(i, \mathbf{Z}) \to \Omega K(i+1, \mathbf{Z})$, and $\mathbf{BU}=\{BU_i\}$ with $BU_i = \mathbf{Z} \times BU$ for i even, and $BU_i = U$ for i odd; the adjoint of $\Sigma BU_i \to BU_{i+1}$ corresponds to $\mathbf{Z} \times BU \simeq \Omega U$ respectively $U \simeq \Omega(\mathbf{Z} \times BU)$. One writes $K^i(X)$ for the cohomology theory associated to $\mathbf{BU}$, and $K_i(X)$ for the corresponding homology theory. Thus

$$\text{for } i \text{ even:} \quad K_i(X) = \pi_i(X_+ \wedge \mathbf{BU}); K^i(X) = [X, \mathbf{Z} \times BU]$$

and

$$\text{for } i \text{ odd:} \quad K_i(X) = \pi_i(X_+ \wedge \mathbf{BU}); K^i(X) = [X, U].$$

For an arbitrary compact space Z the group $K^0(Z) := [Z, \mathbf{Z} \times BU]$ agrees with the *Grothendieck group of complex vector bundles over* Z; similarly for $K^1(Z)$. The *K-homology* groups $K_i(Z)$ admit also a geometric interpretation, as certain *bordism* groups (cf. M. Jakob [38]).

Because $BU \simeq BSU \times K(\mathbf{Z}, 2)$ and $U \simeq SU \times K(\mathbf{Z}, 1)$ with BSU 3-connected and SU 2-connected, one has for a CW-complex X

$$K^0(X) \cong H^0(X; \mathbf{Z}) \oplus H^2(X; \mathbf{Z}), \quad \text{if } \dim X \leq 3$$

and, if X is connected,

$$K^1(X) \cong H^1(X; \mathbf{Z}) \cong \mathrm{Hom}(\pi_1(X), \mathbf{Z}), \quad \text{if } \dim X \leq 2.$$

It is easy to check that these isomorphisms are natural isomorphisms of *groups.*

Since for an arbitrary CW-complex X there is a natural *universal coefficient sequence* (cf. [91])

$$0 \to \mathrm{Ext}(K_{i-1}(X), \mathbf{Z}) \to K^i(X) \to \mathrm{Hom}(K_i(X), \mathbf{Z}) \to 0,$$

the following holds by comparing the K-theory universal coefficient sequence with that for ordinary (co)homology.

Lemma A.4.1. *If X is a two dimensional CW-complex, then there are natural isomorphisms*

$$K_0(X) \cong H_0(X;\mathbf{Z}) \oplus H_2(X;\mathbf{Z}); \quad K_1(X) \cong H_1(X;\mathbf{Z}).$$

All spectra are rationally equivalent to Eilenberg–Mac Lane spectra, and therefore there is, for an arbitrary homology theory $\mathbf{h}$ given by a spectrum, a *generalized Chern character*

$$Ch_n : \mathbf{h}_n(X) \to \bigoplus_{i+j=n} H_i(X;\mathbf{h}_j(\{*\}) \otimes \mathbf{Q}$$

which is an isomorphism upon tensoring with $\mathbf{Q}$. In case of K-theory this translates into the following.

Lemma A.4.2. *For an arbitrary CW-complex X one has natural isomorphisms*

$$K_0(X) \otimes \mathbf{Q} \cong \bigoplus_i H_{2i}(X;\mathbf{Q}); \quad K_1(X) \otimes \mathbf{Q} \cong \bigoplus_i H_{2i+1}(X;\mathbf{Q}).$$

The previous two lemmas are special cases of the following.

Exercise: It is well-known that the classifying space BU has k-invariants $k^{2i} = 0$ and k^{2i+1} of order $(i-1)!$, see [73]. Use this fact to show that if X is a $2N$-dimensional CW-complex then

$$(K_0(X) \oplus K_1(X)) \otimes \mathbf{Z}[1/N!] \cong \bigoplus_i H_i(X;\mathbf{Z}[1/N!]).$$

Author's address:
DEPT. OF MATHEMATICS, ETH ZÜRICH, SWITZERLAND
E-mail address: mislin@math.ethz.ch

Bibliography

[1] C. Anantharaman-Delaroche, J. Renault. *Amenable groupoids.* With a foreword by Georges Skandalis and Appendix B by E. Germain. Monographies de L'Enseignement Mathématique, 36, Geneva, 2000. 196 pp. ISBN 2-940264-01-5.

[2] M. F. Atiyah. *Elliptic operators, discrete groups and von Neumann algebras.* Astérisque 32/33 *(1976)*, 43–72.

[3] M. F. Atiyah. *K-Theory.* Notes by D. W. Anderson. Second edition. Advanced Book Classics. Addison-Wesley Publishing Company, *1989*, ISBN 0-201-09394-4.

[4] M. F. Atiyah. *Global theory of elliptic operators.* Proc. int. Symp. on Funct. Anal., Univ. of Tokyo Press, Tokyo *1969*, 21–30.

[5] M. F. Atiyah, I. M. Singer. *The index of elliptic operators. I.* Ann. of Math. (2) **87**, *(1968)*, 484–530.

[6] H. Bass. *Euler characteristics and characters of discrete groups.* Invent. Math. **35**, *(1976)*, 155–196.

[7] P. Baum, A. Connes. *K-theory for Lie groups and foliations.* Enseign. Math. (2) **46** *(2000)*, no. 1-2, 3–42.

[8] P. Baum, A. Connes. *Chern characters for discrete group.* A fête of topology, Academic Press *(1988)*, 163–232.

[9] P. Baum, A. Connes, N. Higson. *Classifying space for proper actions and K-theory of group C^*-algebras.* C^*-algebras: 1943–1993 (San-Antonio, TX, 1993), 240–291, Contemp. Math., **167**, Amer. Math. Soc., Providence, RI, *(1994)*.

[10] P. Baum, R. G. Douglas. *K-homology and index theory.* Proc. Symposia in Pure Math. Vol XXXVIII, Amer. Math. Soc., Providence, Rhode Island *(1982)*, 117–173.

[11] C. Beguin, H. Bettaieb, A. Valette. *K-Theory for C^*-algebras of one-relator group.* K-theory **16**, 277–298, *(1999).*

[12] M.E.B. Bekka, M. Cowling, P. de la Harpe. *Some groups whose reduced C^*-algebra is simple.* Publ. Math. I.H.E.S. **80**,*(1994)*, 117–134.

[13] H. Bettaieb, M. Matthey, A. Valette. *Low-dimensional group homology and the Baum-Connes assembly map.* Preprint *1999.*

[14] H. Bettaieb, A. Valette. *Sur le groupe K_1 des C^*-algèbres réduites de groupes discrets.* C. R. Acad. Sci. Paris Sér. I Math. **322** *(1996)*, no. 10, 925–928.

[15] I. Chatterji. *Property (RD) for uniform lattices in products of rank one Lie groups with some rank two Lie groups.* Preprint 2001, to appear in Geom. Dedicata.

[16] P. -A. Cherix, M. Cowling, P. Jolissaint, P. Julg, A. Valette. *Groups with the Haagerup property (Gromov's a-T-menability).* Birkäuser. Progress in Math. 197, 2001.

[17] A. Connes. *Noncommutative differential geometry.* Inst. Hautes Études Sci. Publ. Math. **62** *(1985)*, 257–360.

[18] A. Connes. *Noncommutative geometry.* Academic Press, *1994.*

[19] T. Delzant. *Sur l'anneau d'un groupe hyperbolique.* C. R. Acad. Sci. Paris Sér. I Math. **324** *(1997)*, no. 4, 381–384.

[20] T. tom Dieck. *Transformation Groups.* De Gruyter Studies in Math. **8** *(1987).*

[21] A. Dold. *Lectures on Algebraic Topology.* Springer Grundlehren Vol. 200, Springer Verlag *1972.*

[22] C. H. Dowker. *Topology of metric complexes.* Amer. J. Math. **75** *(1952)*, 555–577.

[23] B. Eckmann. *Cyclic homology of groups and the Bass conjecture.* Comment. Math. Helv. **61** *(1986)*, no. 2, 193–202.

[24] B. Eckmann. *Idempotents in a complex group algebra, projective modules and the von Neumann algebra.* Arch. Math. (Basel) **76** *(2001)*, no. 4, 241–249.

[25] G. A. Elliott, T. Natsume. *A Bott periodicity map for crossed products of C^*-algebras by discrete groups.* K-Theory **1** *(1987)*, no. 4, 423–435.

[26] D. R. Farkas, R. L. Snider. *K_0 and Noetherian group rings.* J. Algebra **42** *(1976)*, no. 1, 192–198.

[27] S. C. Ferry, A. Ranicki, J. Rosenberg (eds.) *Novikov conjectures, index theorems and rigidity*, London Math. Soc. Lect. Note Ser. **226**, Cambridge U.P. *(1995)*.

[28] M. Gromov. *Asymptotic invariants of infinite groups.* Geometric group theory, Vol. 2 (Sussex, 1991), 1–295, London Math. Soc. Lecture Note Ser., **182**, Cambridge Univ. Press, *1993*.

[29] M. Gromov. *Spaces and questions.* Preprint, Nov. *1999*.

[30] U. Haagerup. *An example of a nonnuclear C^*-algebra which has the metric approximation property.* Invent. Math. **50** *1979*, 279–293.

[31] P. de la Harpe. *Groupes hyperboliques, algèbres d'opérateurs et un théorème de Jolissaint.* C. R. Acad. Sci. Paris Sér. I **307** *(1988)*, 771–774.

[32] P. de la Harpe, A. Valette. *La propriété (T) de Kazhdan pour les groupes localement compacts.* Astérisque **175** Soc. Math. de France *1989*.

[33] N. Higson. *Bivariant K-theory and the Novikov conjecture.* Geom. Funct. Anal. **10** *(2000)*, no. 3, 563–581.

[34] N. Higson, G. Kasparov. *Operator K-theory for groups which act properly and isometrically on Hilbert space.* Invent. Math. **144** *(2001)*, no. 1, 23–74.

[35] N. Higson, V. Lafforgue, G. Skandalis *Counterexamples to the Baum-Connes conjecture.* Preprint February 2001.

[36] N. Higson, J. Roe. *Analytic K-homology.* Oxford Mathematical Monographs, Oxford University Press, 2000. ISBN: 0-19-851176-0

[37] D. Husemoller. *Fibre bundles.* Graduate Texts in Mathematics, 20. Springer-Verlag, New York, 1994. xx+353 pp. ISBN: 0-387-94087-1

[38] M. Jakob, *A bordism-type description of homology*, Manuscripta Math. **96** (1998), 67–80.

[39] R. Ji. *Smooth dense subalgebras of reduced group C^*-algebras, Schwartz cohomology of groups, and cyclic cohomology.* J. Funct. Anal. **107** *(1992)*, 1–33.

[40] P. Jolissaint. *Rapidly decreasing functions in reduced C^*-algebras of groups.* Trans. Amer. Math. Soc. **317** *(1990)*, 167–196.

[41] P. Julg. *C^*-algèbres associées à des complexes simpliciaux.* C. R. Acad. Sci. Paris Sér. I Math. 308 *(1989)*, no. **4**, 97–100.

[42] P. Julg. *Travaux de N. Higson et G. Kasparov sur la conjecture de Baum-Connes.* Séminaire Bourbaki. Vol. 1997/98. Astérisque No. 252 (1998), Exp. No. 841, 4, 151–183.

[43] P. Julg, G. Kasparov. *L'anneau $KK_G(\mathbf{C}, \mathbf{C})$ pour $G = \mathrm{SU}(n, 1)$.* J. Reine Angew. Math. **463** *(1995)*, 99–152.

[44] P. Julg, A. Valette. *K-theoretic amenability of $SL_2(\mathbf{Q}_p)$, and the action on the associated tree.* Journ. Funct. Anal. **58** *(1984)*, 194-215.

[45] G. Kasparov. *K-theory, group C*-algebras, and higher signatures (Conspectus).* Novikov conjectures, index theorems and rigidity, Vol. 1 (Oberwolfach, 1993), 101–146, London Math. Soc. Lecture Note Ser., 226, Cambridge Univ. Press, 1995.

[46] G. Kasparov. *Hilbert C*-modules: theorems of Stinespring and Voiculescu.* J. Operator Theory **4** *(1980)*, 133–150.

[47] G. Kasparov. *The operator K-functor and extensions of C*-algebras.* Math. USSR Izv. **16** *(1981)*, 513–672.

[48] G. Kasparov. *The index of invariant elliptic operators, K-theory, and Lie group representations.* Dokl. Akad. Nauk. USSR, vol. **268**, *(1983)*, 533–537.

[49] G. Kasparov. *Lorenz groups: K-Theory of unitary representations and crossed products.* Sov. Math. Dockl. **29** *(1984)*, 256–260.

[50] G. Kasparov. *Equivariant KK-theory and the Novikov conjecture.* Invent. Math. **91** *(1988)*, 147–201.

[51] G. Kasparov, G. Skandalis. *Groups acting on buildings, operator K-theory, and Novikov's conjecture.* K-Theory **4** *(1991)*, no. 4, 303–337.

[52] G. Kasparov, G. Skandalis. *Groupes "boliques" et conjecture de Novikov.* C. R. Acad. Sci. Paris Sér. I Math. **319** *(1994)*, no. 8, 815–820.

[53] N. Keswani. *Homotopy invariance of relative eta-invariants and C*-algebra K-theory.* Electron. Res. Announc. Amer. Math. Soc. **4** *(1998)*, 18–26 (electronic).

[54] J. L. Koszul. *Lectures on groups of transformations.* Tata Institute of Fundamental Research, Bombay, *1965.*

[55] D. Kucerovsky. *The Baum-Connes assembly map in the unbounded picture.* Appendix to [88].

[56] P. Kropholler, G. Mislin. *Groups acting on finite-dimensional spaces with finite stabilizers.* Comment. Math. Helv. **73** *(1998)*, no. 1, 122–136.

[57] V. Lafforgue. *Une démonstration de la conjecture de Baum-Connes pour les groupes réductifs sur un corps p-adique et pour certains groupes discrets possédant la propriété (T).* C. R. Acad. Sci. Paris Sér. I Math. 327 *(1998)*, no.**5**, 439–444.

[58] V. Lafforgue. *A proof of property (RD) for discrete cocompact subgroups of* $SL_3(\mathbf{R})$ *and* $SL_3(\mathbf{C})$. Journal of Lie Theory **10**, *(2000)*, 255–267.

[59] V. Lafforgue. *K-Théorie bivariante pour les algèbres de Banach, groupoïdes et conjecture de Baum-Connes.* Preprint, to appear in Inventiones Math.

[60] E. C. Lance. *Hilbert C*-modules, a toolkit for operator algebraists.* London Math. Soc. Lecture Notes Ser. **210** *(1995).*

[61] A. Lubotzky, S. Mozes, M. S. Raghunathan. *Cyclic subgroups of exponential growth and metrics on discrete groups* C. R. Acad. Sci. Paris, t.317, Sér. I,p.735–740, *(1993).*

[62] Z. Marciniak. *Cyclic homology and idempotents in group rings.* Transformation groups, Poznań 1985, 253–257, Lecture Notes in Math., **1217**, Springer, Berlin-New York, *1986.*

[63] M. Matthey. *K-theories, C*-algebras and assembly maps.* Thesis, Neuchâtel, 2000.

[64] J. P. May, *Equivariant Homotopy and Cohomology Theory* AMS Regional Conference Series in Mathematics **91**, *(1996).*

[65] I. Mineyev, G. Yu. *The Baum-Connes conjecture for hyperbolic groups.* MSRI Preprint, *2001.*

[66] A. S. Miščenko. *Infinite-dimensional representations of discrete groups, and higher signatures. (Russian)* Izv. Akad. Nauk SSSR Ser. Mat. **38** *(1974)*, 81–106.

[67] G. Mislin. *Equivariant K-homology of the classifying space for proper actions.* Proceedings of the Euro Summer School on proper group actions, Barcelona September 2001.

[68] T. Natsume. *The Baum-Connes conjecture, the commutator theorem and Rieffel projections.* C. R. Math. Rep. Acad. Sci. Canad. **1** *(1988)*, 13–18.

[69] S. P. Novikov. *Algebraic construction and properties of Hermitian analogs of K-theory over rings with involution from the viewpoint of the Hamiltonian formalism. Applications to differential topology and the theory of characteristic classes.* Izv. Akad. Nauk SSSR Ser. Mat. **34** *(1970)*, 253–288, 475–500; English translation, Math. USSR-Izv. **4** *(1970)*, 257–292, 479–505.

[70] H. Oyono-Oyono. *La conjecture de Baum-Connes pour les groupes agissant sur les arbres.* C. R. Acad. Sci. Paris Sér. I Math. **326** *(1998)*, no.7, 799–804.

[71] H. Oyono-Oyono. *Baum-Connes conjecture and extensions.* J. Reine Angew. Math. **532** *(2001)*, 133–149.

[72] D. S. Passman. *The algebraic structure of group rings.* Robert E. Krieger Publishing Co., Inc., Melbourne, Fla., *1985.* ISBN:0-89874-789-9.

[73] F. Peterson. *Some remarks on Chern classes.* Annals of Math. **69** *(1959)*, 414–420.

[74] M. Pimsner, D. Voiculescu. *Exact sequences for K-groups and EXT-groups of certain cross-product C^*-algebras.* Journal of Operator Theory **4** *(1980)*, 93–118.

[75] L. S. Pontryagin. *Topological groups.* Gordon and Breach Science Publishers, Inc., New York-London-Paris *1966.*

[76] J. Ramagge, G. Robertson, T. Steger. *A Haagerup inequality for $\widetilde{A}_1 \times \widetilde{A}_1$ and $\widetilde{A}_2$ buildings.* Geom. Funct. Anal. 8 *(1998)*, no. **4**, 702–731.

[77] J. Rosenberg. *C^*-algebras, positive scalar curvature, and the Novikov conjecture.* Inst. Hautes Études Sci. Publ. Math. No. **58** *(1983)*, 197–212.

[78] G. Skandalis. *Une notion de nucléarité en K-théorie.* K-theory **1** *(1987)*, 549–573.

[79] G. Skandalis. *Kasparov's bivariant K-theory and applications.* Expo. math. **9** *(1991)*, 193–250.

[80] G. Skandalis. *Progrès récents sur la conjecture de Baum-Connes. Contribution de Vincent Lafforgue.* Sém. Bourbaki, nov. *1999*, exposé **869**.

[81] E. H. Spanier. *Algebraic Topology.* McGraw-Hill 1966.

[82] S. Stolz. *Concordance classes of positive scalar curvature metrics.* Preprint *1998.*

[83] R. G. Swan. *Vector bundles and projective modules.* Trans. Amer. Math. Soc. **(105)** *1962*, 264–277.

[84] R. M. Switzer. *Algebraic Topology – Homotopy and Homology.* Grundlehren Springer Vol. 212 (1973).

[85] J. L. Taylor. *Banach algebras and topology.* Algebra in Analysis, J. H. Williamson (editor), Academic Press, *1975.*

[86] J. L. Tu. *The Baum-Connes conjecture and discrete group actions on trees.* K-Theory **17** no.4 *(1999)*, 303–318.

[87] A. Valette. *The conjecture of idempotents: a survey of the C^*-algebraic approach* Bull. Soc. Math. Belg., vol **XLI**, *1989*, 485–521.

[88] A. Valette. *The Baum-Connes assembly map.* Proceedings of the Euro Summer School on proper group actions, Barcelona September 2001.

[89] F. Warner. *Foundations of differentiable manifolds.* Graduate Texts in Mathematics, **94**. Springer-Verlag, New York-Berlin, *1983.* ISBN 0-387-90894-3.

[90] N. E. Wegge-Olsen. *K-Theory and C^*-algebras. A friendly approach.* Oxford Science Publications. The Clarendon Press, Oxford University Press, New York, *1993* ISBN 0-19-859694-4.

[91] Z. Yoshimura. *A note on complex K-theory of infinite CW-complexes.* Journal of the Mathematical Society of Japan **26**, No.2 *(1974)*, 289–295.

[92] R. J. Zimmer. *Ergodic theory and semisimple groups.* Monographs in Mathematics, **81**. Birkhäuser Verlag, Basel-Boston, Mass., *1984.* ISBN 3-7643-3184-4.

[93] B. Zimmermann. *Surfaces and the second homology of a group.* Monatsh. Math. **104** *(1987)*, no. 3, 247–253.

Index